PLANETOLOGY

PLANETOLOGY

UNLOCKING THE SECRETS OF THE SOLAR SYSTEM

TOM JONES AND ELLEN STOFAN

NATIONAL GEOGRAPHIC

WASHINGTON, D.C.

CONTENTS

 EARTH   NEPTUNE

EARTH'S MOON  SATURN

JUPITER DISTANT GALAXY

 MARS  URANUS

 MERCURY   VENUS

About this book:
Planet icons located in the upper left-hand corner of the
pages indicate the planets under consideration for that page.
Cross-references, also located in the upper left, direct read-
ers to additional information.

Introduction

THE TWO OF US met for the first time in 1988, as we completed our graduate studies in planetary science. NASA, through its Graduate Research Fellowship Program, brought us together in Washington, D.C., as we presented the findings of our dissertation projects. Tom's project was a search for water on dark, nearly unaltered asteroids far out in the main asteroid belt. Ellen's was a study of volcanoes on Venus. Our meeting started a friendship and working relationship that's lasted 20 trips around the sun.

We met again four years later, this time to collaborate on an exciting "Mission to Planet Earth." Ellen was a planetary scientist working at NASA's Jet Propulsion Laboratory. Tom had completed astronaut training in Houston at NASA's Johnson Space Center, and had just been assigned as a mission specialist on shuttle mission STS-59, aboard *Endeavour*. The shuttle would carry into orbit on two flights the Space Radar Laboratory, a cutting-edge suite of instruments designed to reveal Earth's surface in startling, unprecedented detail.

Ellen was a key member of the Space Radar Lab science team, planning the radar observations and coordinating the activities of both ground investigators and the shuttle crew. We participated in field training at several sites around the globe to prepare for the orbital science campaign. During 1994, *Endeavour* flew twice with its multi-frequency radar and air pollution sensor; Space Radar Lab 1 and 2 (STS-59 and STS-68, with Tom aboard both missions) returned more than 100 terabytes (100,000 gigabytes) of digital imagery, revealing new details of the dynamic surface of our planet. The successful flights were a vivid example of the way NASA has applied planetary remote sensing technology to our study of Earth. The Space Radar Lab, for example, used an advanced imager evolved from the Magellan probe's radar, an instrument that in 1990-94 had mapped the surface of Venus.

Since that successful collaboration, we have continued our work in planetary science. Ellen is an active field and laboratory researcher, studying volcanic activity on Earth, Venus, Mars, and Titan. Her work as a science team member on the Mars Express and Cassini missions puts her at the center of our investigations of the red planet and Saturn's enigmatic largest moon, Titan.

Tom is still interested in asteroids, especially Near Earth Objects, those asteroids and comets that approach and sometimes strike Earth. He assists the Association of Space Explorers in their international efforts to develop, through the United Nations, plans to prevent future devastating impacts.

Tom has had an incomparable experience looking back at Earth on four shuttle flights. Ellen has been among the first to see—using spacecraft radar—the cloud-shrouded surfaces of Venus, and now one of Saturn's moons, Titan. Together, we decided to combine our complementary voices and backgrounds for a fresh look at the processes that shape the worlds of our solar system. We set out to share our scientific enthusiasm and mutual fascination for both the methods and results of space exploration. Our hope is to give the reader an exciting overview of the latest discoveries in our planetary system.

The space age is just over half a century old; we began to see the planets close up just four decades ago. As we developed instruments to explore the solar system, we simultaneously turned the same technologies toward the study of our own planet. Spacecraft not only track the paths of hurricanes and provide real-time navigation for our cars, they provide fundamental scientific information that helps us understand the past, present, and future of Earth. From space, we have detected and monitored the holes in the ozone layer at Earth's poles. That knowledge enabled us to ban the use of ozone-depleting chemicals, and now the holes are slowly closing.

One of the unexpected and long-lasting results of the Apollo moon landing program was the startlingly beautiful sight of earthrise above the lunar horizon. We saw Earth as a unique oasis in space, home to our human species, and the only known harbor for life anywhere in the solar system. By breaking the bonds of Earth, our view changed from a local or regional perspective to a global one. We also understood in a visceral way that ours is just one of many planets in our solar system, one of many solar systems in our galaxy, one of many galaxies in the universe.

Since Apollo we have examined, with our robot explorers, the faces of all but one of the planets in our solar system (with a visit to minor planet Pluto on tap in 2015). In the last decade, an exciting new era of solar system exploration has returned a flood of scientific data and imagery from Mercury, Venus, the moon, Mars, Jupiter, Saturn, and their families of satellites. Those crisp views of other worlds and the latest discoveries about the forces that shaped their surfaces connect those planets more tightly to Earth, the most dynamic of the planets and the textbook for understanding what we see out there.

What we've gained by studying the solar system is a new appreciation of our home world. A doctor with only a single patient might grow quite familiar with his condition, but without examining others, she would learn little about the normal state of humans, or the nature, variety, and progression of diseases. Similarly, planetary scientists need a suite of worlds to examine in order to develop an accurate understanding of how a planet works. Planetologists in this age of space exploration have the wild good luck, for example, of being able to compare crustal faults on the icy moons of Uranus to faults on Mars, and to tectonic faults on Earth. We can compare an active volcanic eruption on Jupiter's moon Io to one occurring on Earth. We can even compare how the wind builds sand dunes on four different planets and moons.

These "field experiments" in planetology constantly refine and deepen our understanding of how our own world ticks. As the poet T. S. Eliot said: "We shall not cease from exploration, and the end of all our exploring will be to arrive where we started and know the place for the first time." This is the fundamental concept of planetology—we study all worlds, so that in the end, we will come to understand our own.

In *Planetology,* we will share with you a variety of views of the worlds that make up the solar system. We have Earth images from the shuttle's Space Radar Lab missions and astronaut handheld photography. We add highlights from the last two decades of planetary spacecraft images and the latest discoveries from the far reaches of the solar system to illustrate the power of comparative planetology.

As scientists, our most rewarding experiences have come from personally exploring our own Earth: both our starting point and the key to understanding distant worlds. As scientists and humans with a sense of wonder, we will enjoy sharing with you our field impressions of our home planet. Yes, we will travel the solar system, but our voyage of exploration in these pages will always bring us back to the third planet, this island Earth.

TOM JONES & ELLEN STOFAN
JUNE 2008

Previous Pages:

i: Earth and Moon, seen from Mars, 88 million miles away, by the Mars Reconnaissance Orbiter in October, 2007.

ii-iii: Glowing knots of gas ejected by a dying star, called the Dumbbell Nebula, M27, 1200 light years from Earth in the Milky Way; imaged by the Hubble Space Telescope in November, 2001.

iv-v: Our home planet, Earth, as seen from the space shuttle

EARTH: KEY TO KNOWING OTHER WORLDS

Understanding our home planet Earth is essential for observing and interpreting other worlds in our solar system—and beyond.

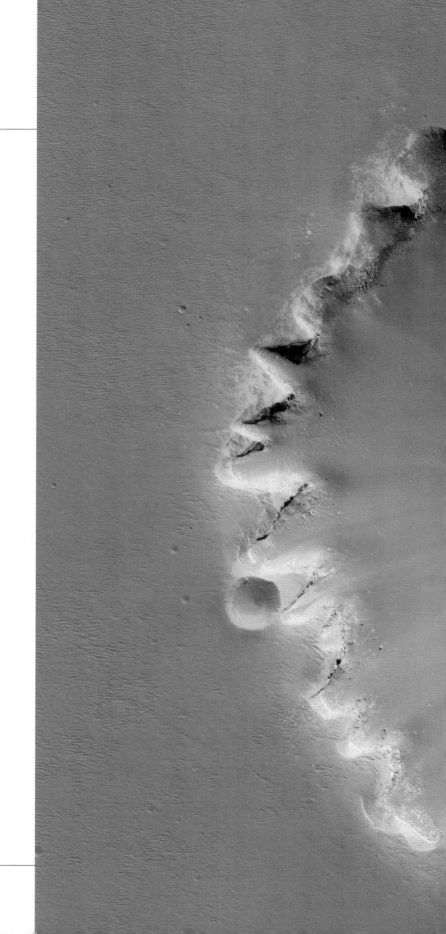

What Is Planetology?

SEE ALSO: *Similar Yet Unique* 14

VER SINCE THEY DISCERNED that the wandering planets were different from the stars of the firmament, our ancestors wondered if there were other worlds like our own. For millennia no answer was possible, but since the invention of the telescope scientists have debated whether the blurry, dimly glimpsed surfaces of the planets bore any relationship to landscapes on Earth. Only with the advent of the space age, and the launch of robotic emissaries and finally astronauts beyond the atmosphere to other worlds, did close-up images reveal the true surfaces of alien worlds.

To their surprise, planetary scientists discovered that as strange as the lunar and Martian landscapes turned out to be, they in fact bore characteristics familiar to Earth geologists. Although early spacecraft images showed that the surfaces of Mars and the moon were dominated by impact craters, scientists soon found that traces of ancient impacts are hardly uncommon on Earth. Later probes and the Apollo missions found other parallels to our own world, including massive volcanoes, uplifted mountain ranges, and fault scarps running hundreds of miles. The moon has been resurfaced by immense floods of lava, like those found in Siberia, India, and the Pacific Northwest of the United States. Mars harbors huge volcanoes, thick polar ice caps, and, like Earth, extensive networks of streams, valleys, and flood channels. Our scientific study of Earth has lasted barely more than three centuries; our close-up scrutiny of the planets, just four decades. Now, we are even finding worlds circling other stars.

Earth's geologic features taught us what to look for on these other worlds, but our growing understanding of the sun's planetary family has in turn yielded new insights into the forces shaping the surface of our own globe. This concept of using other worlds to test our ideas of how Earth functions is called comparative planetology. By learning how external and internal geologic processes work on a variety of worlds—with different temperatures, mineral composition, and gravitational fields—we can better understand the details of how those forces operate on our Earth.

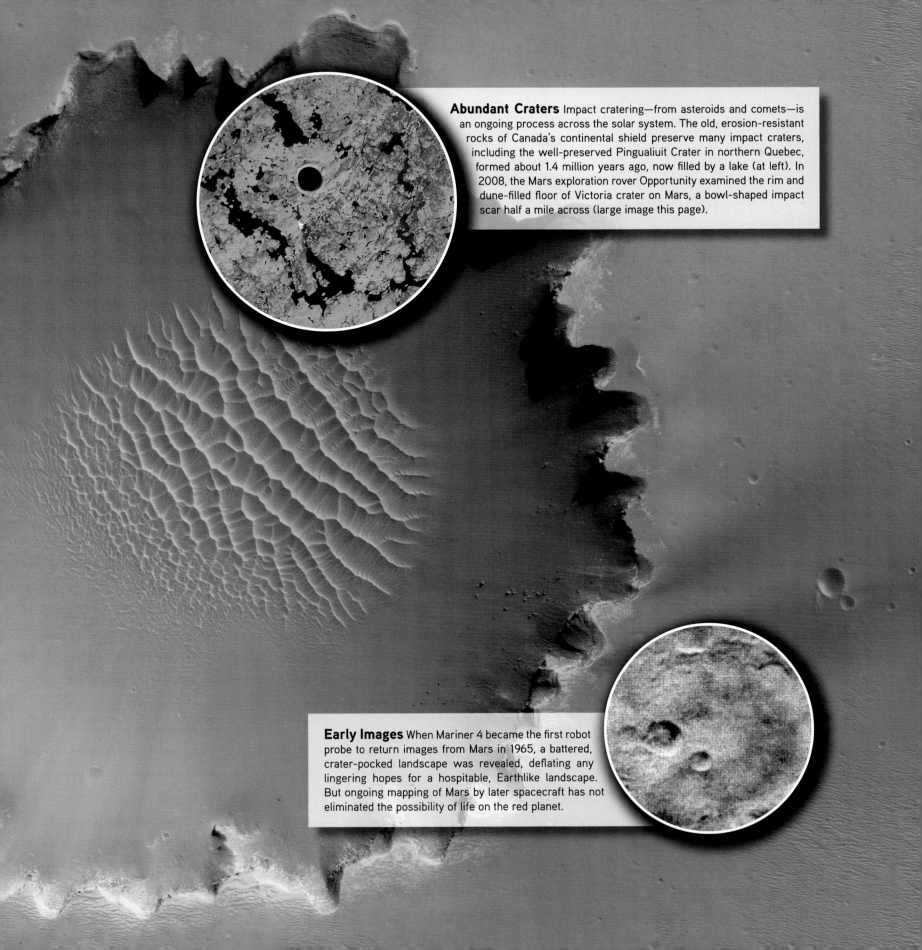

Abundant Craters Impact cratering—from asteroids and comets—is an ongoing process across the solar system. The old, erosion-resistant rocks of Canada's continental shield preserve many impact craters, including the well-preserved Pingualiuit Crater in northern Quebec, formed about 1.4 million years ago, now filled by a lake (at left). In 2008, the Mars exploration rover Opportunity examined the rim and dune-filled floor of Victoria crater on Mars, a bowl-shaped impact scar half a mile across (large image this page).

Early Images When Mariner 4 became the first robot probe to return images from Mars in 1965, a battered, crater-pocked landscape was revealed, deflating any lingering hopes for a hospitable, Earthlike landscape. But ongoing mapping of Mars by later spacecraft has not eliminated the possibility of life on the red planet.

Planet Earth: *An Introduction*

THE STORY OF PLANETOLOGY starts here, on planet Earth. Our world is third from the sun, the largest of the terrestrial planets, those four rocky worlds sunward of Jupiter. We glide along our orbit about the sun at about 66,000 miles an hour, 93 million miles from our star. We're at just the right distance for water to exist as a solid, liquid, and gas. Earth has a solid iron core, a liquid iron outer core, a semisolid or "plastic" mantle that makes up the bulk of the planet's mass, and a thin, brittle crust.

Both Earth's crust and the biosphere cloaking it in life are the products of 4.5 billion years of planetary evolution. Our active world has churned and metamorphosed since its formation, influenced by the never ceasing work of internal and external forces. From within and without, Earth's surface has been roiled by colliding plates of rock buckling into mountain ranges, continuous eruptions of lavas forming new crust, and the hammer blows of celestial impacts. Our planet's crust, about 18 miles thick under the continents and just 3 miles thick under the oceans, buckles and breaks because of movements of the mantle beneath. The mantle's rocks, stretching 1,800 miles from core to crust, flow like warm molasses, heated by radioactive elements such as uranium, thorium, and potassium. Those convection currents push and pull the crust, lifting mountains, generating fractures (faults), fueling volcanoes, and sliding crustal plates across the face of the globe. These internal forces not only shift the crust and make volcanoes and mountain ranges but also create new oceanic crustal material along mid-ocean rifts, like the Mid-Atlantic Ridge. At other plate boundaries, called subduction zones, older crust plunges back into the mantle for melting and recycling. On a timescale of hundreds of millions of years, Earth renews its ocean floors. Except for a few slivers of ancient crust plastered to the edges of the continents, few traces remain of oceanic crust older than 200 million years.

On the lighter, older, continental plates, running water, grinding ice, and scouring winds have combined to attack every structure that Earth's internal tectonic forces could heave skyward. Where erosion has spared the ancient cores of the continents, we find ancient remnants of the original continental crust: Some rocks in Greenland date from 4.2 billion years ago. Over the long, 4.5-billion-year history of our planet, colliding continental plates have raised mountain ranges again and again,

4.6 BILLION YEARS AGO
Solar system is formed from a
giant cloud of dust and gas

4.5 BILLION YEARS AGO
Planets are formed

4.5 BILLION YEARS AGO
Earth beings to cool;
atmosphere forms

3.6 BILLION YEARS AGO
One-celled organisms arise

EARTH: **KEY TO KNOWING OTHER WORLDS**

yet erosion has always triumphed, laying low the highest peaks and carrying the fragments away for deposition as new sedimentary rocks.

Rising and evolving with that same dynamic surface, life has established itself from the deepest mid-ocean vents to the tiniest pores in the crust, from the frozen rocks and ice of Antarctica to the near-boiling hot springs of Yellowstone. Life owns Earth. But life's dominance of Earth has not been inevitable; our planet is sometimes downright hostile to its presence here.

Earth is but one of eight major worlds circling the sun, and the history of life on Earth is strongly tied to the processes of the solar system. Life's energy derives almost exclusively from the sun's light and heat. Life is constantly affected by the power of natural disasters like earthquakes, hurricanes, and volcanic explosions. Such events punctuate the long and active history of our planet.

On a dynamic world like Earth, change is a constant. We are compelled by our survival instincts to understand the forces that drive that change, in hopes of adapting to new conditions, or, as with the threat from cosmic collisions, using our technology to prevent a catastrophe we cannot survive.

A NEW EARTH

The Apollo 8 crew took this photograph of our home world on December 24, 1968, on the first mission to orbit the moon. Earth-circling satellites had previously taken global images of our world, but with Apollo 8's photo for the first time we humans saw Earth as an oasis in a silent, indifferent cosmos: our hospitable "blue marble" floating invitingly above the moon's blasted, lifeless surface. The astronauts of Apollo 8 studied maps of the lunar surface obtained by the Lunar Orbiter series of robot spacecraft. They trained with field geologists on Earth in order to identify the landforms—craters, mountain chains, impact basins, crustal rifts—they would see from orbit. Later Apollo crews participated in extensive field exercises to apply the lessons of Earth geology to the six lunar landing expeditions. Astronauts traveled to lava fields in Iceland, Hawaii, Arizona, and Idaho. They hiked the steep interior slopes of impact scars like Meteor Crater in Arizona. They learned to analyze regional geology and identify volcanic and impact-formed rocks on sight. Today, astronauts receive basic geology training through classroom lectures and field orientation trips. NASA is now setting up training programs for new astronaut classes to prepare future crews for a return to the moon—and field exploration there—sometime after 2020.

1 BILLION YEARS AGO
Multicelled organisms arise

470 MILLION YEARS AGO
Land plants arise; fishes are dominant life form on Earth

250 MILLION YEARS AGO
Dinosaurs dominate Earth

210 MILLION YEARS AGO
First mammals appear

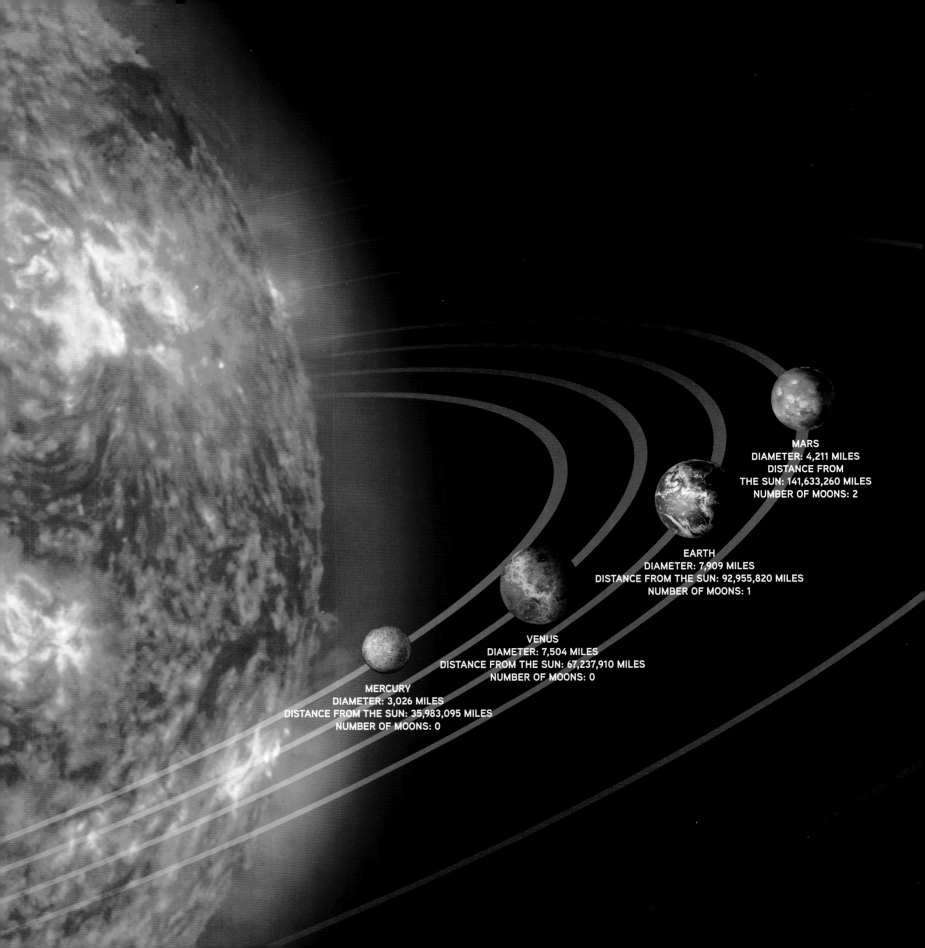

MARS
DIAMETER: 4,211 MILES
DISTANCE FROM
THE SUN: 141,633,260 MILES
NUMBER OF MOONS: 2

EARTH
DIAMETER: 7,909 MILES
DISTANCE FROM THE SUN: 92,955,820 MILES
NUMBER OF MOONS: 1

VENUS
DIAMETER: 7,504 MILES
DISTANCE FROM THE SUN: 67,237,910 MILES
NUMBER OF MOONS: 0

MERCURY
DIAMETER: 3,026 MILES
DISTANCE FROM THE SUN: 35,983,095 MILES
NUMBER OF MOONS: 0

NEPTUNE
DIAMETER: 30,710 MILES
DISTANCE FROM THE SUN: 2,795,084,800 MILES
NUMBER OF MOONS: 13

URANUS
DIAMETER: 31,693 MILES
DISTANCE FROM THE SUN: 1,783,939,400 MILES
NUMBER OF MOONS: 27

SATURN
DIAMETER: 74,732 MILES
DISTANCE FROM THE SUN: 885,904,700 MILES
NUMBER OF MOONS: 60

JUPITER
DIAMETER: 88,650 MILES
DISTANCE FROM THE SUN: 483,682,810 MILES
NUMBER OF MOONS: 63

FEATURE:
THE SOLAR SYSTEM

Our star, Sol, shepherds a family of eight major planets. Closest to the sun are the rocky terrestrial planets; all but Mercury have kept substantial atmospheres. The main asteroid belt, a ring of debris left over from planetary formation, lies between Mars and Jupiter. Next come the gas giants, Jupiter and Saturn, which swept up large amounts of hydrogen and helium from the solar nebula during their formation. Jupiter is more than ten times Earth's diameter. Uranus and Neptune are also gaseous planets, but are only a third Jupiter's size. Pluto, once considered the ninth planet, is just one of many small, icy bodies inhabiting the outer rim of the solar system. It joins a few large asteroids in the category of minor planets.

Select Moons *of the Solar System*

O F THE SOLAR SYSTEM'S planets, all but two—Mercury and Venus— have satellites, or moons. Natural satellites are thought to have different origins. As the massive gas giant planets formed from the solar nebula, they were surrounded by protosatellite disks, in much the same way as a newborn star forms amid a protoplanetary disk. The swirling cloud of dust and gas gave rise to bodies big enough to be worlds in their own right, surrounding Jupiter, Saturn, and even Uranus with their own miniature solar systems. The best examples are Jupiter's four large Galilean satellites, and Saturn's Titan, Rhea, Dione, Iapetus, Tethys, and Enceladus. Other satellites are about the size of asteroids and comets and were probably captured into orbit after near collisions with a planet: Phobos (opposite page) and Deimos around Mars are probable examples. The odds are small that an asteroid could be captured into a stable, long-term orbit, but as the outer planets formed, their gravity kicked thousands of small bodies into planet-crossing orbits. No doubt a few survived a close encounter and were captured, braked by a planet's extended atmosphere or protosatellite disk. At least one moon—our own—probably formed in the aftermath of a giant collision, as did Neptune's Triton. These violent collisions blasted material from both the parent planet and impactor into orbit, where the superheated gas, dust, and debris coalesced into a

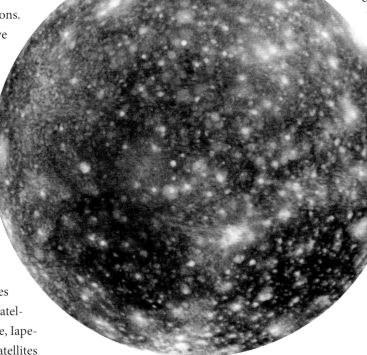

With a diameter of more than 2,976 miles, Callisto is Jupiter's second largest moon, just slightly smaller than Mercury. Its icy crust is densly pockmarked by impact craters, indicating that its crust may be up to 4 billion years old, recording the solar system's last spurt of heavy asteroid and comet bombardment.

moon. Neptune, which rotates on its side, about 90 degrees from the other planets' axes, probably was struck by a giant impact that knocked the planet askew and ejected the material that formed Triton. Computer models of solar system formation suggest that Mercury and Venus, which don't have moons today, were also likely struck by large asteroids or comets. But these impacts either did not eject material that formed a moon (instead falling back onto the planet) or the primitive moon that did form was dragged from orbit by tides and absorbed once more by its parent.

Most small satellites quickly leaked away their internal heat to the cold of space, but a handful were large enough to retain the warmth from radioactive elements and generate internal geological activity. Volcanoes erupted, lavas flowed, crust buckled and twisted into mountain ranges, and whole blocks of crust lifted or sank along tectonic faults. Io, constantly flexed by the mammoth tides raised by Jupiter, is so volcanically active that its lava-and-sulfur surface is the youngest in the solar system. The sheer variety of moons provides us with many more "laboratories" to test our theories of comparative planetology. Smaller satellites, whose interiors have cooled and hence ended internally generated geological activity, act as excellent recorders of solar system history. Their stable, solid surfaces began to record the scars of smaller impacts, enabling us to read the story of bombardment episodes throughout the solar system.

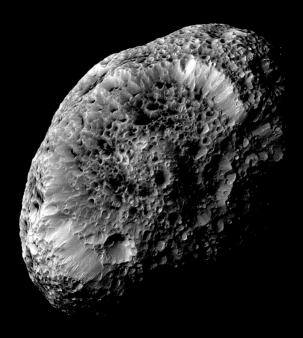

Hyperion, just 180 miles across, is a shattered fragment of the demolition derby underway among Saturn's 50-plus moons. A collision has ripped away half of the moon, exposing its frothy, ice-rich interior for inspection by Cassini in 2005. Its low density implies that it is made mostly of fluffy, porous ice.

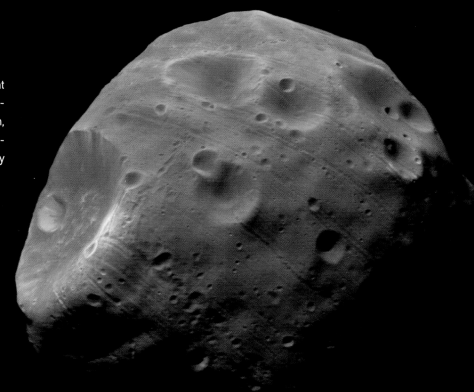

The battered, dust-cloaked surface of Phobos, a moon of Mars, was captured in crisp detail by the Mars Express camera in 2008. The impact that dug giant Stickney crater (left), more than six miles across, may have also etched the enigmatic grooves scoring this potato-shaped 13-mile-long moon.

OBSERVING EARTH

Geologists have always sought the high ground to assess the lay of the land, and aerial photography gave scientists a welcome vantage point. Artificial satellites took that concept to new heights, providing reconnaissance to the military (the U.S. Corona series was the first, about 1960), and then giving science a global perspective with Landsat and others. On California's northwest coast lies San Francisco Bay, bisected by the San Andreas Fault, a tectonic fault where the Pacific and the North American crustal plates grind past each other. Modern San Francisco stands astride the fault, which in 1906 lurched more than 20 feet, destroying more than 28,000 buildings and killing more than 3,000 people. This large image pictures the financial district, Embarcadero, and Oakland Bay Bridge, seen from the IKONOS satellite. To observe this geologically important region, scientists have used not just orbital photography but portions of the electromagnetic spectrum our eyes cannot see. Visible light can show us details of street plans, vegetation patterns, and surface geology. But other wavelengths can reveal the type and health of vegetation, delineate land-use patterns, and trace the unstable tectonic underpinnings of this vulnerable region. In the same way, planetary scientists examine a planet's surface using a range of wavelengths, looking for different rock types (lava, crustal bedrock, water-bearing sediments, ice buried under dust) or revealing landforms shrouded by haze or clouds.

INTERNATIONAL SPACE STATION

Urban areas, vegetation, and silt flowing into San Francisco Bay are visible in this medium-resolution digital image taken by astronatuts aboard the International Space Station.

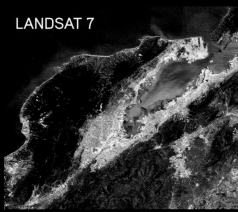

LANDSAT 7

Landsat's digital scanners see the same region in different wavelengths. Comparing brightness in colors reveals vegetation and urban development.

SAR SATELLITE

Radar uses its own microwave "light" to illuminate a scene, through darkness or clouds. Spaceborne radar can detect minute ground movements associated with earthquakes.

Similar, Yet Unique

SEE ALSO: *Martian Canals* 110

SCIENTISTS QUICKLY APPLIED OUR TOOLS for observing Earth to unravel the secrets of other worlds. The first rush of results came from the 1960s space race, as NASA scrutinized the moon to find safe landing sites for Apollo crews. Mars received the next round of attention, as it held out promise that conditions there might once have been hospitable to life. Through the 1970s, a flood of imagery came back from lunar and Mars orbiters and landers. Geologists used their terrestrial experience to interpret what they were seeing. Although the moon's craters and volcanic features had some terrestrial parallels, it was Mars that amazed planetary geologists with its similarities to Earth. Yet the red planet exhibited its own unique spin on its varied landscapes.

The volcanoes were immediately recognizable as shield volcanoes, with easy slopes and summit calderas typical of basaltic volcanoes on Earth. But the Martian examples were gigantic: The largest, Olympus Mons, was as big as the state of Arizona, dwarfing Mauna Loa, Hawaii, Earth's most massive shield volcano.

Mariner 9 revealed spectacular canyons and networks of what appeared to be stream valleys. On Earth, those channels meant running water, supplied by rainfall. Was the same true for Mars? With an array of sophisticated cameras now circling Mars, scientists are taking a close look for answers. NASA's Mars Reconnaissance Orbiter's HiRISE camera imaged these gullies near Gorgonum Chaos, in Mars's southern hemisphere (right). Carved into the inner wall of an impact crater, these gullies seem to have been formed by flowing fluids, probably water. The gullies appear to originate around a sequence of rocky layers near the crater rim.

Boulders litter the channels, as if the flow removed finer sediments and left the larger, heavier rocks behind. The small, interwoven channels in each gully suggest repeated episodes of flowing liquid. By comparing the depth and contours of these valleys on Mars to those on Earth (inset), we can learn whether rainfall, or another process, best explains their formation. One candidate for the source of channels on Mars is infrequent outbursts of subsurface water from a layer just below the crater's lip.

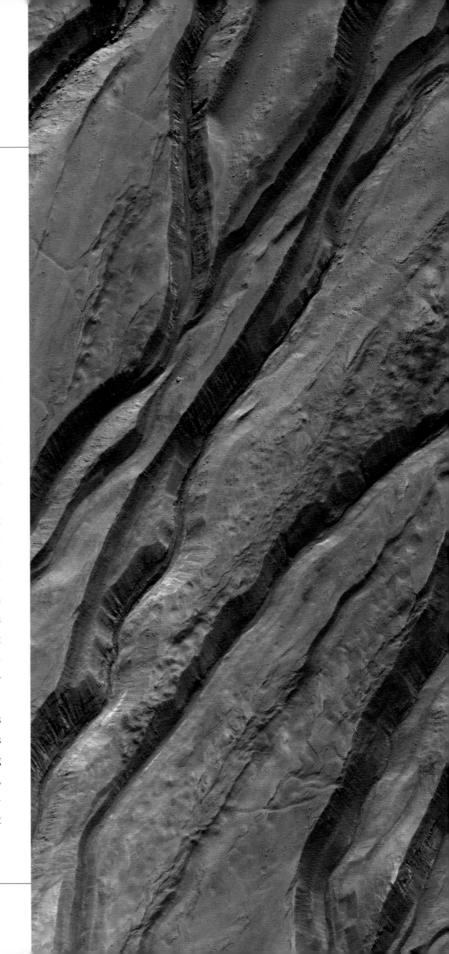

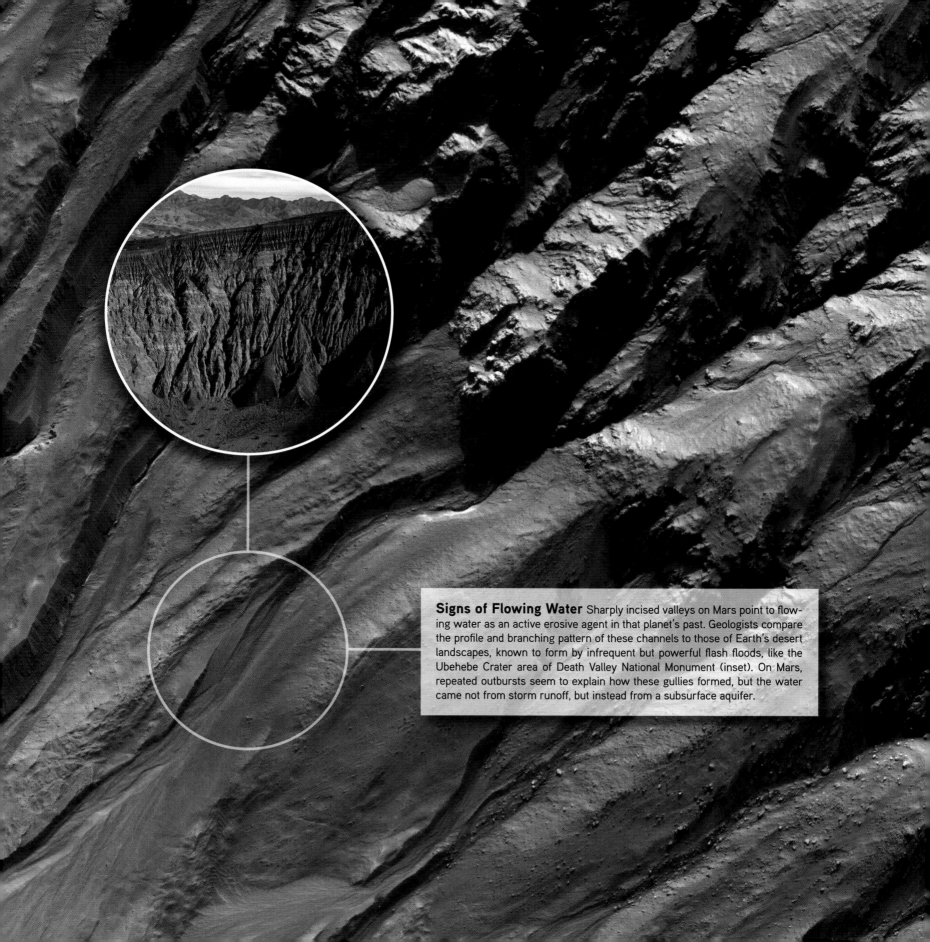

Signs of Flowing Water Sharply incised valleys on Mars point to flowing water as an active erosive agent in that planet's past. Geologists compare the profile and branching pattern of these channels to those of Earth's desert landscapes, known to form by infrequent but powerful flash floods, like the Ubehebe Crater area of Death Valley National Monument (inset). On Mars, repeated outbursts seem to explain how these gullies formed, but the water came not from storm runoff, but instead from a subsurface aquifer.

EXPEDITION EARTH

On my four field expeditions to Earth orbit, totaling 53 days, I took full advantage of my orbital vantage point to explore our exquisitely beautiful planet. My two Space Radar Lab (SRL) flights, STS-59 and STS-68, both aboard shuttle *Endeavour*, produced a fresh new look at the active surface below. On the first mission, in April 1994, our state-of-the-art imager took 939 radar observations during 65 hours of operation. Space Radar Lab 1 scanned 5.4 percent of Earth's surface, filling 166 cassette tapes with 47 trillion bits of radar imagery. Printed out on paper, those images would fill 20,000 volumes of an encyclopedia documenting the changes, both natural and man-made, affecting Earth's surface. My crew of astronaut observers was busy, too, returning 11,000 photos of our science targets. Flying with SRL-2, in October 1994, our team captured another 80 hours of radar imaging, enough to fill a 65-foot-high stack of compact discs, and another 13,000 frames of science photography. The photography and radar imagery provided Ellen and our international science team with years of research material to investigate our dynamic world. Seeing Earth from above is an incomparable privilege. No science experience of mine can match the satisfaction I felt after three weeks observing our world through *Endeavour*'s big flight deck windows; I could hardly tear myself away from the view. As we explore the solar system, our reference point will always be the knowledge drawn from further study of our home planet.   —*Tom Jones*

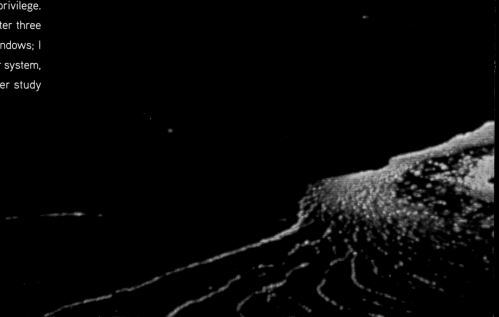

In Training Author and astronaut Tom Jones trained in Hawaii for his 1993 Space Radar Lab mission by trekking Kilauea's lava flows with volcanologists, members of the science team comparing "ground truth" with orbital radar images. Planetary geologist and author Ellen Stofan still performs field studies at Sicily's Mount Etna volcano.

SEE ALSO: *Where Do We Look?* 176

Observing the Planets

GALILEO FIRST TURNED his new telescopes toward the planets in 1609, glimpsing the crescent Venus, the four big moons of Jupiter, and the starkly desolate face of the moon. But even as late as 1965, ground-based telescopes could give us only vague hints of surface markings on Mars, and the best images of the moon were too blurred by our atmosphere to search out safe landing zones for astronauts. It took an extension of our own eyes by robot spacecraft to open up the planetary frontier.

The first robotic probes to visit our planetary neighbors glimpsed landscapes seared by scorching temperatures on Venus or scarred by devastating bombardment on Mars. Flybys of the moon, Venus, and Mars soon gave way to more sophisticated orbiters to map nearby worlds. Five lunar orbiters charted the moon for Apollo. In 1971 Mariner 9 revealed a complex and ancient Mars. In the 1970s and '80s, the Pioneers and Voyagers made the first flyby reconnaissance of Jupiter, Saturn, and the distant outer solar system. Voyagers 1 and 2 revealed for the first time the faces of the Jovian and Saturnian satellites, mere pinpoints of light to us until the space age.

By the 1990s, planetary exploration reached a new level of sophistication. Magellan used radar to map Venus's cloud-shrouded surface. Our return to Mars with landers (Pathfinder and Sojourner, its midget rover) and orbiting spacecraft (Mars Global Surveyor, Mars Odyssey, Mars Express, Mars Reconnaissance Orbiter) makes the red planet our most studied—but still puzzling—neighbor.

Initially, the prospects for life elsewhere in our solar system appeared bleak, seemingly confirmed by the sterile samples returned from the moon by Apollo's astronauts. But the revolution in planetary science spurred by the Voyager, Galileo, and Cassini spacecraft and the recent flood of information from Mars has raised interest in the new field of astrobiology. Did the conditions for life ever exist elsewhere in the solar system, or outside our own planetary family? Again, our detailed exploration of Earth's surface provides the baseline to compare with new findings from other worlds.

The dynamic, 4.5-billion-year evolution of our planet has long since erased the earliest chapters of life's story here. But in ancient landscapes across the solar system, we may find clues as to how life may have originated. On Earth, the interactions between the atmosphere, crust, oceans, ice caps, and life are incredibly complex. By looking carefully at older, less active worlds—for organic compounds on comets or asteroids, at hot springs near the dormant volcanoes of Mars, in the subsurface ocean on Jupiter's moon Europa—we may make important comparisons to help understand the processes that gave rise to our habitable Earth.

Radar telescopes like Arecibo (below) in Puerto Rico can tease out fine surface details, such as roughness and soil properties, that complement the global mapping by robot spacecraft. Arecibo's narrow radar beam can light up the shadowed floors of polar craters on the moon, looking for ice deposits that might one day support a lunar base.

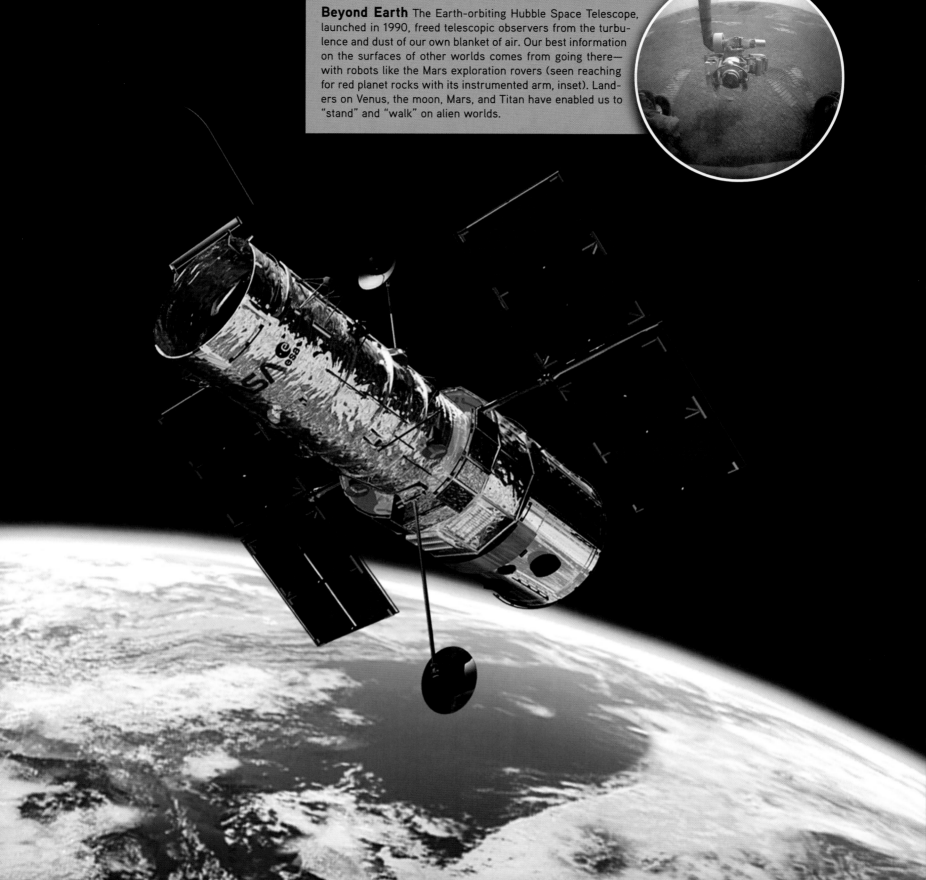

Beyond Earth The Earth-orbiting Hubble Space Telescope, launched in 1990, freed telescopic observers from the turbulence and dust of our own blanket of air. Our best information on the surfaces of other worlds comes from going there—with robots like the Mars exploration rovers (seen reaching for red planet rocks with its instrumented arm, inset). Landers on Venus, the moon, Mars, and Titan have enabled us to "stand" and "walk" on alien worlds.

Time Line of Space Exploration

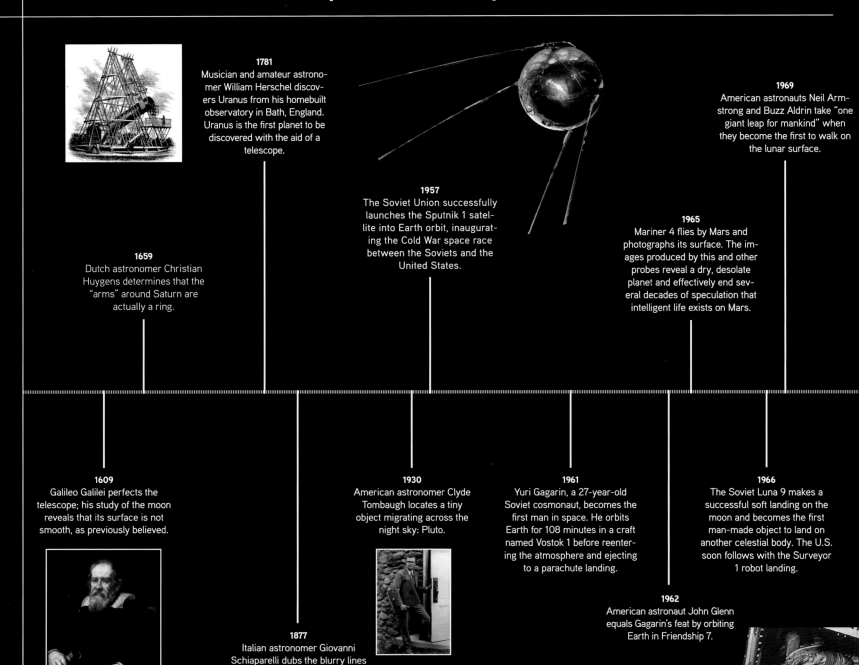

1781
Musician and amateur astronomer William Herschel discovers Uranus from his homebuilt observatory in Bath, England. Uranus is the first planet to be discovered with the aid of a telescope.

1957
The Soviet Union successfully launches the Sputnik 1 satellite into Earth orbit, inaugurating the Cold War space race between the Soviets and the United States.

1969
American astronauts Neil Armstrong and Buzz Aldrin take "one giant leap for mankind" when they become the first to walk on the lunar surface.

1659
Dutch astronomer Christian Huygens determines that the "arms" around Saturn are actually a ring.

1965
Mariner 4 flies by Mars and photographs its surface. The images produced by this and other probes reveal a dry, desolate planet and effectively end several decades of speculation that intelligent life exists on Mars.

1609
Galileo Galilei perfects the telescope; his study of the moon reveals that its surface is not smooth, as previously believed.

1930
American astronomer Clyde Tombaugh locates a tiny object migrating across the night sky: Pluto.

1961
Yuri Gagarin, a 27-year-old Soviet cosmonaut, becomes the first man in space. He orbits Earth for 108 minutes in a craft named Vostok 1 before reentering the atmosphere and ejecting to a parachute landing.

1966
The Soviet Luna 9 makes a successful soft landing on the moon and becomes the first man-made object to land on another celestial body. The U.S. soon follows with the Surveyor 1 robot landing.

1877
Italian astronomer Giovanni Schiaparelli dubs the blurry lines on Mars's surface "canali"— giving rise to speculation that there may be life on Mars.

1962
American astronaut John Glenn equals Gagarin's feat by orbiting Earth in Friendship 7.

1979
Voyager 2 flies by Jupiter and spots a volcano erupting on Io, one of the moons Galileo discovered.

2004
Opportunity, one of NASA's Mars rovers, collects rock samples suggesting that parts of Mars may have once been under water.

1973
NASA launches Mariner 10, the first mission to Mercury. It will make three flybys of this planet closest to the sun, measuring its mass and magnetic field.

1989
NASA dispatches the Galileo spacecraft to Jupiter. Galileo sends a probe hurtling into Jupiter's cloud tops and discovers a subsurface liquid ocean on Europa.

1997
The Mars Pathfinder lands on the red planet to explore and photograph it.

2007
NASA launches the Dawn spacecraft on an eight-year mission to understand the origins of the solar system. Dawn will orbit Vesta and Ceres, two of the largest asteroids in the asteroid belt, and gather data on them.

1975
NASA launches Viking 1 and Viking 2, to map and explore Mars, to take extensive soil samples, and to monitor wind and temperature.

1990
The space shuttle *Discovery* carries the Hubble Space Telescope into space. In the following years, Hubble greatly advances our knowledge of the universe.

2005
The orbiting Cassini half of the Cassini-Huygens spacecraft photographs Saturn's moon Titan and finds seas which are likely filled not with water but with liquid ethane and methane. The Huygens probe detached and landed on the surface of Titan.

1989
NASA launches Magellan to Venus. The probe maps most of the planet's surface, charting volcanoes and "continents."

1997
NASA launches Cassini-Huygens toward Saturn, where it will measure its physical dimensions and its chemical and atmospheric components.

SEE ALSO: *Ice Volcanoes* 96

Putting It All Together

O F COURSE, EARTH IS THE STARTING POINT. Our ability to visit and investigate in detail terrestrial landforms and geologic processes is tremendously valuable. Through field observation and experiment, earth scientists have built theories or hypotheses about how volcanoes erupt, earthquakes rumble, and glaciers advance and retreat. Today, researchers apply those models to the features seen on other planets; if those geological features can be explained by our model, then our understanding is reinforced. If solar system reality does not fit our Earth-generated theory, then the model needs revision. Geologists head back into the field for more data, and theorists go back to reassess their approach to explaining fundamental forces like impacts, volcanism, and erosion. By incorporating the data and imagery from robotic probes and astronaut expeditions into the study of our planet, we not only learn more about the solar system but also about our own Earth.

The solar system is still surprising us. Recently the joint American and European probe Cassini, in orbit around Saturn, delivered two stunning findings. First, Cassini images in March 2006 showed that Saturn's tiny and frigid moon Enceladus, which seemed to be been frozen in rigor mortis, exhibited fresh surface cracks and incredibly, geysers of liquid water spraying ice droplets into space. The source of the geysers? Internal heat generated by tidal stretching from Saturn's immense gravity.

Then, the following August, Cassini bounced a stream of radar signals off the surface of Saturn's largest, haze-shrouded moon, Titan. What appeared in the echoed digital images were dark, smooth splotches on that satellite's rock-hard, water ice surface. The evidence suggests that these dark markings are Titanian lakes—ponds of liquid methane, rained from the hydrocarbon-rich smog in the nitrogen atmosphere. The methane lakes are yet more evidence of Titan's complex chemistry, which may preserve the early building blocks that under warmer conditions, as on a young Earth, might have given rise to life. Those chemical precursors have long since been destroyed on dynamic Earth; the Titan discoveries may enable us to travel back in space and time to sample those compounds.

Intriguing Moons Saturn's smog-shrouded moon, Titan, (large gold image) home to lakes of supercold methane, looms behind the rings of its giant parent; the battered moonlet Epimethus, 72 miles across, floats near the rings (top center) in this 2006 Cassini image. Titan's complex organic chemistry may harbor clues to the origins of life in the solar system. Its surprisingly active sister satellite, Enceladus (inset), jets water vapor and organic compounds into space through cracks in its icy surface.

OUR CATASTROPHIC PAST

Dramatic evidence of our planet's violent history, Barringer Meteor Crater in Arizona is a nearly mile-wide reminder of the asteroid impacts that have helped shape Earth's surface.

Marked by Catastrophe: *Our Scarred Planet*

THIRTY-FIVE MILLION YEARS AGO, the Atlantic Ocean washed North America's shores 93 miles farther west of today's coastline, the result of higher global sea levels. Rain forests choked the slopes of the Appalachian Mountains, thriving in a warm climate even more humid than a muggy mid-Atlantic summer. Suddenly, in the hazy blue tropical sky, a brilliant white light backlit banks of clouds, followed seconds later by a blinding flash and the bang of a thousand thunderclaps. An asteroid or comet, perhaps three miles in diameter, slammed into the shallow sea covering the continental shelf, generating a fireball and blast that ignited or pulverized everything within hundreds of miles. When the shock wave had faded, the giant tsunamis had subsided, and the scalding debris had rained down, a yawning crater scarred the limestones and deeper granites underlying the roiling sea. Resembling the multiringed craters of the moon, the undersea impact scar lay sterile for a thousand years or more. Sediments and the microscopic shells of plankton eventually filled and buried the crater, and it lay undetected at the mouth of today's Chesapeake Bay until geologists drilled down to it in the early 1990s. The ancient scar affects us even today: Its slumped and broken rocks channel salty brine into the freshwater wells of modern Virginia's coastal residents.

Sudden catastrophic events such as meteor impacts, massive volcanic eruptions, and earthquakes have had a profound impact on the history of our solar system, changing the landscape, ushering in new eras of evolution, and even creating our nearest neighbor—the moon. Understanding the nature of these events is crucial to building a picture of our past—and assuring our own survival.

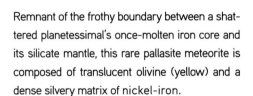

Remnant of the frothy boundary between a shattered planetessimal's once-molten iron core and its silicate mantle, this rare pallasite meteorite is composed of translucent olivine (yellow) and a dense silvery matrix of nickel-iron.

By far the most sudden and destructive process affecting Earth and its neighbors is that of cosmic impacts. The inner planets have been struck repeatedly—and disastrously—throughout their history by small remnants from the formation of the solar system: asteroids and comets. Asteroids are the remnants of millions of planetesimals—mini-planets that coalesced between Mars and Jupiter in the early days of the solar system. But Jupiter's rapid growth stirred the gravitational pot, scattering many of these small bodies into the rest of the solar system, preventing the formation of a larger planet in what is now the asteroid belt. Those errant asteroids, and the shattered fragments of their collisions, careened through the orbits of the inner planets and Jupiter's satellites, peppering their surfaces with thousands of impacts.

Simultaneously, icy planetesimals formed in the region of Uranus and Neptune—comets—were hurled by the gravity of those growing planets into the inner solar system, adding to the carnage. Careful tallying of the craters on Mercury, the moon, and Mars—our planetary neighbors—tells us how many collisions have occurred in the past and gives scientists an approximate count of the number of objects that were loose during the first half billion years of our history. Samples collected during the Apollo program's lunar missions tell us the timing of that bombardment.

Over time, the planets swept up the rogue planetesimals or kicked them entirely out of the solar system. But hundreds of thousands of objects, ranging from minor planets like Ceres and Vesta to small fragments no larger than a house, still populate the asteroid belt. Collisions and Jupiter's continuing gravitational influence send a steady stream of asteroids sunward, and swarms of asteroids and dormant comets approach or cross Earth's orbit.

Crater Cousins In the young solar system, planets and moons were heavily battered by asteroid and comet impacts, a process that continues sporadically today. The collision that formed the Spider crater (inset) produced these distinctive fractures in Mercury's crust. On Earth, weathering and crustal movement has destroyed most evidence of this violent past. In northern Canada, however, ancient rocks preserve the 210-million-year-old Manicouagan crater (large image), which stretches more than 60 miles across. Rivers here may also trace impact-generated fault lines.

Slow & Steady— *Most of the Time*

"THE PRESENT IS THE KEY TO THE PAST"— these famous words of geologist James Hutton, an 18th-century father of that discipline, sum up our thinking on how Earth's surface became what we see today. Hutton proposed that the geological processes that sculpted Earth operate on very long timescales, compared to our brief life spans, and change the world around us in a very gradual, steady, almost unnoticeable way.

The stability of our Earth is a comfortable concept, given the rapid technological and societal changes that swirl around us. Seen from orbit, our world seems peaceful and unchanging, with only shifting clouds and storms subtly altering the astronaut's view. To be sure, space station crews, aloft for six months or more, do see seasonal changes: the poleward retreat of sea ice in the spring, the rush of green returning to Midwest farms in early summer, the fading of deciduous forests to autumn's rusty brown, and the spreading mountain snows of deepening winter. By contrast, spotting geological change from space over scales of a few months is nearly impossible.

But a careful observer notices, even from 200 miles up, small but distinct shifts in the landscape. Volcanoes fume, spew gouts of lava, and smear the atmosphere with plumes of ash. Saharan dust storms swirl west across the Atlantic, depositing dust in Florida and the Caribbean. Landslides and fresh mudflows scar the steep slopes of mountain valleys, and seasonal floods push rivers through newly carved channels.

These are the changes visible on human timescales, but in the last century geologists have discovered evidence that occasionally, sudden and dramatic shifts have greatly altered our familiar landscapes. Devastating floods of lava and explosive volcanic eruptions have buried entire regions under blankets of fresh rock or ash. Earthquakes have shifted massive blocks of crust along fault lines, destroying cities in the process. And impacts from the heavens have repeatedly scarred our planet, devastated its surface, and shifted the very course of life on Earth.

Impacts. Volcanic explosions. Earthquakes. Tsunamis. These destructive geologic processes unfold on very short timescales, but their appearances during most of human history have been rare. The worst of

4 BILLION YEARS AGO
Heavy bombardment from asteroids and comets shapes the early solar system.

250 MILLION YEARS AGO
The largest volcanic eruption in history floods Siberia in lava, possibly causing a mass extinction.

65 MILLION YEARS AGO
A giant meteor crashes into the Yucatán Peninsula, triggering another extinction.

17 MILLION YEARS AGO
Lava flows from the Grande Ronde flood the U.S.'s Northwest with basalt.

these events have spared us of late, but they have left us without many clues about how these dramatic geological processes operate. Over its 4.6-billion-year history, in fact, our active Earth has recycled most of its crust and obliterated nearly all of the direct evidence of how catastrophe has affected its surface. How then can we possibly unravel the story of how these violent forces abruptly shaped the world on which we came to be? Fortunately, most of the other planets and moons of our solar system are not nearly as active as our own. Some of their ancient surfaces preserve evidence of what undoubtedly happened here during Earth's first billion years. To assess how often these violent events occur, how they alter our planet's surface, and what dangers they pose to civilization and life, we look to other worlds.

Over the past five or six million years, the Colorado River has cut slowly and steadily downward through much older sedimentary rock. Given enough time, nearly imperceptible erosion can level mountains or carve a mile-deep chasm.

THE AGE OF EARTH

A pioneer in our search for the true age of Earth was Sir Charles Lyell (1797-1875), a contemporary of Charles Darwin. Lyell realized that if the face of our planet had formed via the gradual process of erosion, its landscapes were far older than Bible-derived estimates: perhaps millions of years old—a then revolutionary concept. His concept of steady change, called uniformitarianism, became the foundation of our understanding of Earth. The true age of Earth was revealed by a technique called radiometric dating, which uses the known rates of radioactive decay of certain elements (like uranium or potassium) to tell us when Earth's geologic clock was set. By measuring the relative abundance in a rock of these key radioactive elements, we can work backward to determine when it was formed. Radiometric dating of meteorites collected on Earth has shown their age to be 4.6 billion years, marking the beginning of planet formation. The oldest moon rocks returned by the Apollo astronauts are between 4.4 and 4.5 billion years old, which gives scientists a firm date for the formation of the moon. Earth's oldest remaining rocks—those that have escaped destruction and recycling by its active erosive and tectonic processes—formed just over 4 billion years ago. Surfaces of the same age on different planets should show the same pattern of crater size and quantity. For example, by carefully comparing the number and size of craters on the lunar plains visited by Apollo 11 with similar regions on Mars, we've dated some of the red planet's surface to 3.6-3.9 billion years old. Its crater-battered southern hemisphere probably dates back 4 billion years, with the northern volcanic plains perhaps only a billion or two years old. Although we don't have samples from Mars itself, aside from a handful of meteorites blasted from its surface, plans for collecting rocks via robot spacecraft and returning them to Earth will one day verify these educated guesses.

A.D. 1556
An earthquake in Shaanxi, China, kills an estimated 830,000 people.

A.D. 1783
Iceland's Laki volcanic eruption lasts eight months, killing 9,000 people and much livestock.

A.D. 1906
A magnitude 9.5 earthquake, the most powerful ever recorded, strikes Chile.

A.D. 2004
An earthquake off Sumatra sets off the Asian Tsunami, killing more than 230,000.

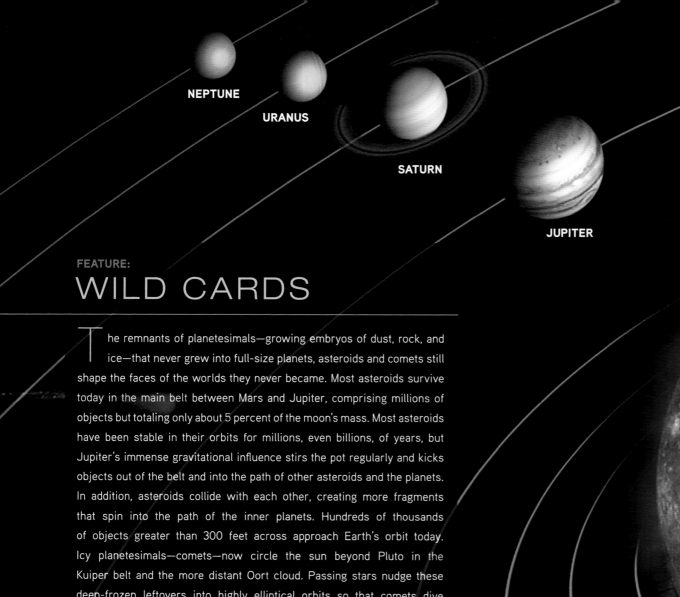

NEPTUNE

URANUS

SATURN

JUPITER

WILD CARDS

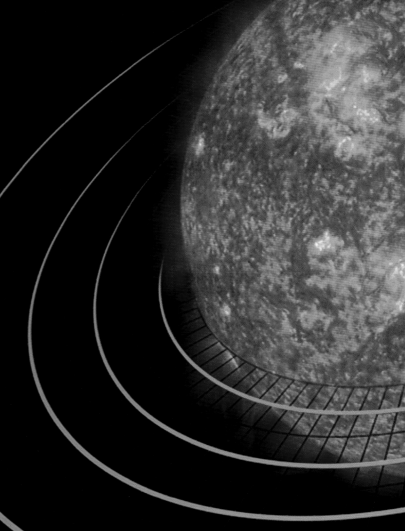

The remnants of planetesimals—growing embryos of dust, rock, and ice—that never grew into full-size planets, asteroids and comets still shape the faces of the worlds they never became. Most asteroids survive today in the main belt between Mars and Jupiter, comprising millions of objects but totaling only about 5 percent of the moon's mass. Most asteroids have been stable in their orbits for millions, even billions, of years, but Jupiter's immense gravitational influence stirs the pot regularly and kicks objects out of the belt and into the path of other asteroids and the planets. In addition, asteroids collide with each other, creating more fragments that spin into the path of the inner planets. Hundreds of thousands of objects greater than 300 feet across approach Earth's orbit today. Icy planetesimals—comets—now circle the sun beyond Pluto in the Kuiper belt and the more distant Oort cloud. Passing stars nudge these deep-frozen leftovers into highly elliptical orbits so that comets dive through the inner solar system at speeds of 25 miles a second or faster. As Earth moves along its orbit, it sweeps up dust from comets and asteroids. These tiny, speeding dust fragments, smaller than sand grains, are incinerated by atmospheric friction and seen as streaks of light in the night sky—meteors. Those rocky shards big enough to survive, strike the ground, and be recovered for study are called meteorites.

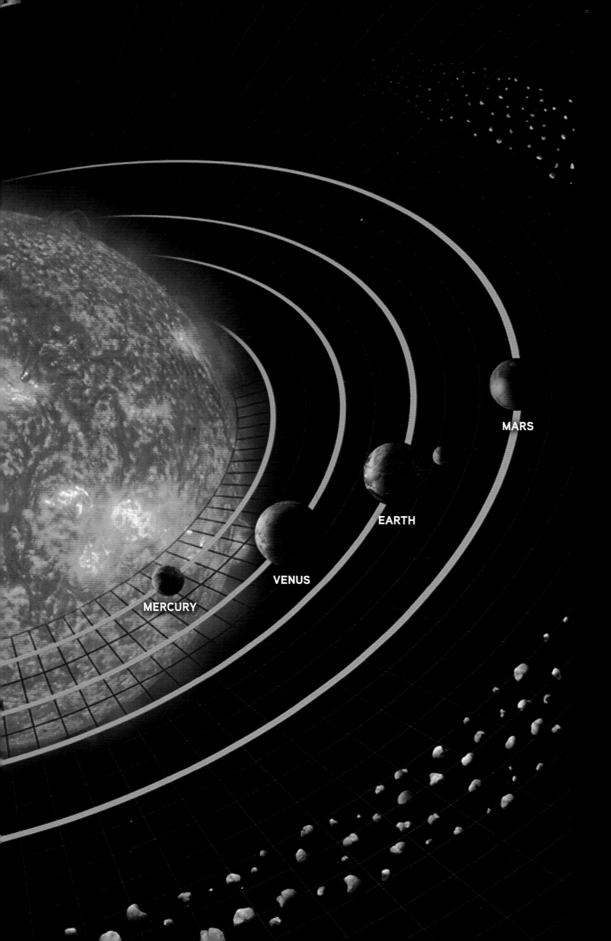

MERCURY

VENUS

EARTH

MARS

COMET

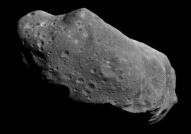

Three key parts of a comet are visible in this image of comet C/2001 Q4: its icy core, the surrounding coma, and the distinctive trailing gases of the tail.

METEOR

The Perseid meteor shower, pictured above, is an annual show of cometary dust burning up in the atmosphere as Earth's orbit crosses through the tail of comet Swift/Tuttle.

ASTEROID

Asteroid 243 Ida, fairly typical in size at 35 miles across, has its own tiny moon, Dactyl, visible as a tiny speck (lower right) in this image captured by the Galileo spacecraft.

SEE ALSO: *Life in the Asteroid Swarm* 40

Deep Impact

IMPACTS FROM asteroids and meteors are the key catastrophic process that has shaped the surfaces of the inner planets and icy moons. When a cosmic body collides with Earth, at velocities of six to seven miles a second or more, it is slowed hardly at all by the atmosphere. The sudden impact generates a shock wave that almost instantaneously converts the energy of motion—kinetic energy—into heat, a heat so intense that it largely vaporizes the impactor. The same shock melts the surrounding native rock and produces a rapidly expanding fireball. A decompression wave following the shock hurls molten and shattered debris away from the impact site in a conical sheet, leaving behind a bowl-shaped cavity—a crater.

In the 1960s, studies of Arizona's Meteor (or Barringer) Crater gave us our first real insight into the physics of impact cratering. In the last 40 years, the study of craters on the moon, Mars, Venus, and the icy satellites of the outer solar system have helped us to refine our models of how craters are formed. Those models predict what effects such high-energy impacts have on a planet's surface, and, for example, on Earth's biosphere.

If the impact is from an object more than a mile or so in diameter, it forms not just a simple bowl-shaped cavity, as at Meteor Crater, but a more complex structure. Debris slumps back into the crater, creating a flat floor and a central uplift, or peak. Larger

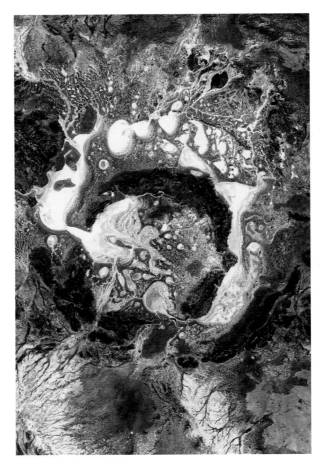

Billions of years of erosion have nearly erased the subtle yet colorful traces of the Shoemaker impact crater in western Australia. The circular pattern of fractured bedrock and minerals created under terrific shock proved the origins of this collision. Dry beds of salt lakes ring the weathered rim of the shallow depression.

impacts create a series of concentric terraces and rings around the original crater rim, as whole blocks of crust are fractured and tilted by the explosion. These complex craters were first studied on the moon, by telescope and robot spacecraft.

Across the solar system, giant asteroid or comet collisions have left traces of huge, multiringed features called impact basins. These huge collisions fractured the crust for hundreds of miles, threw debris thousands of miles from the impact site, and gouged so deeply that magma flooded the low-lying basin to form smooth lava plains. The most familiar examples are the lunar maria (seas), but impact basins are common on Mercury, Mars, and the large icy satellites of Jupiter and Saturn. By studying the characteristic features of these scars, caused by giant impacts on the planets, we learned how to search for evidence of such catastrophic collisions on Earth. Using satellite photos and field studies at ground level, geologists have found traces of giant impact basins preserved on the continental shields, the oldest and most stable rocks of the crust. Vredefort in South Africa and Sudbury in Canada are nearly 200 miles across and approximately two billion years old. These ancient, global catastrophes occurred repeatedly, wiping out all but a few of the simple life forms on Earth and resetting and redirecting life's future course on this planet.

Planetary Scars: Dwarfing most of Earth's impact scars, Herschel crater (upper right of inset) on Saturn's moon Mimas was formed by a hypervelocity collision that nearly shattered the tiny satellite, itself just 243 miles across. Herschel is 80 miles across, with a 3-mile-high rim and a central peak soaring 4 miles above the crater floor. Earth's largest impact basin is Vredefort (large image), 185 miles across, formed two billion years ago when a 6-mile-wide object hit the southern part of present-day Africa.

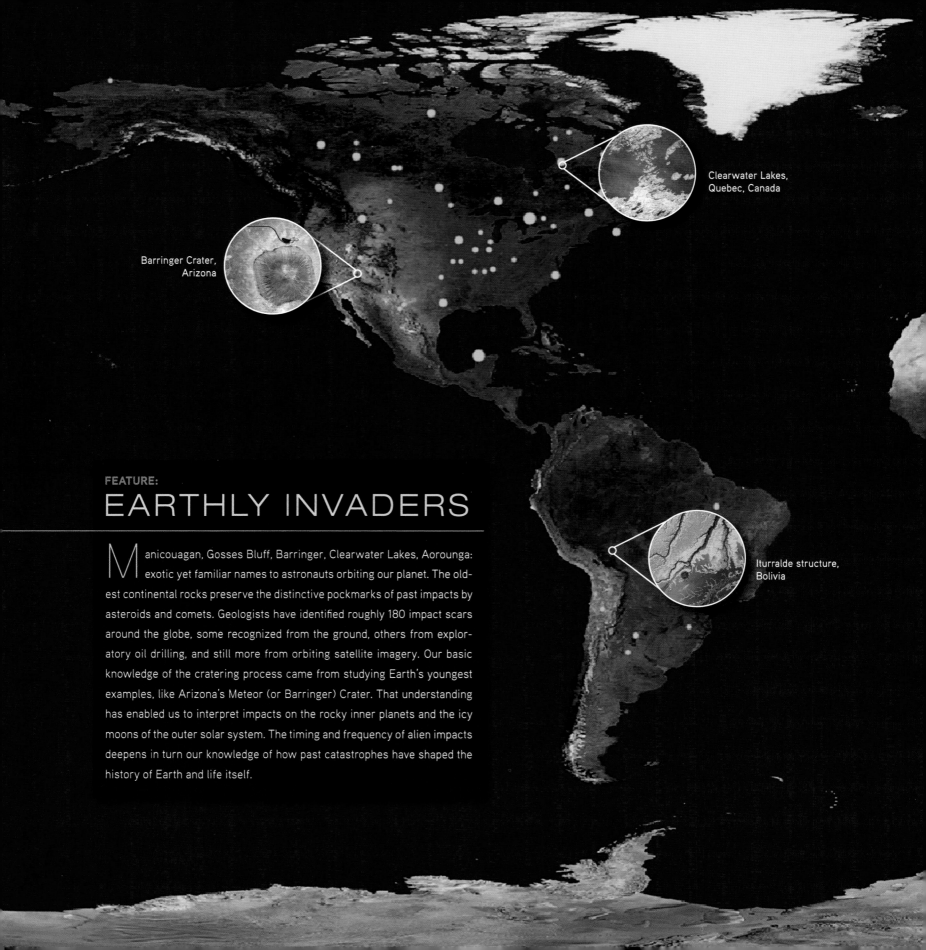

Clearwater Lakes,
Quebec, Canada

Barringer Crater,
Arizona

Iturralde structure,
Bolivia

FEATURE:
EARTHLY INVADERS

M anicouagan, Gosses Bluff, Barringer, Clearwater Lakes, Aorounga: exotic yet familiar names to astronauts orbiting our planet. The oldest continental rocks preserve the distinctive pockmarks of past impacts by asteroids and comets. Geologists have identified roughly 180 impact scars around the globe, some recognized from the ground, others from exploratory oil drilling, and still more from orbiting satellite imagery. Our basic knowledge of the cratering process came from studying Earth's youngest examples, like Arizona's Meteor (or Barringer) Crater. That understanding has enabled us to interpret impacts on the rocky inner planets and the icy moons of the outer solar system. The timing and frequency of alien impacts deepens in turn our knowledge of how past catastrophes have shaped the history of Earth and life itself.

Lake Kaali on
Saaremaa Island, Estonia

Kara-Kul,
Tajikistan

Sahara,
northern Chad

Wolf Creek Impact Crater,
western Australia

Roter Kamm
impact crater, Namibia

Gosses Bluff,
northern Australia

Across the Solar System

S INCE GALILEO FIRST TRAINED his telescope on the heavens in 1609, astronomers have known that the moon was scarred by thousands of craters, but their origin was a subject of argument for centuries. Geologists guessed that the craters were probably volcanic, but a few thought they might have been formed by impacts from the larger cousins of the meteors we see streaking across the night sky. It was not until the early 1960s that geologist and astronomer Eugene Shoemaker, examining the mile-wide Meteor (Barringer) Crater near Winslow, Arizona, proved conclusively that it had been caused by an impact by identifying geologic evidence left by the blast, which was caused by a 50-yard-wide asteroid fragment that struck Earth with a 20-megaton blast. Meteor Crater in Arizona is probably the most famous impact structure on Earth, and despite its age of 50,000 years, its desert location has kept its outlines relatively sharp. NASA's lunar exploration program revealed that most of the moon's craters were likewise formed by

Iapetus, about 905 miles across (above), is one of Saturn's most enigmatic satellites. Covered on one side by black, carbon-rich material, its water-ice crust is a brilliant white. Collisions during its history have created huge impact scars, like this 280-mile-wide basin which overlies an older crater just to its right.

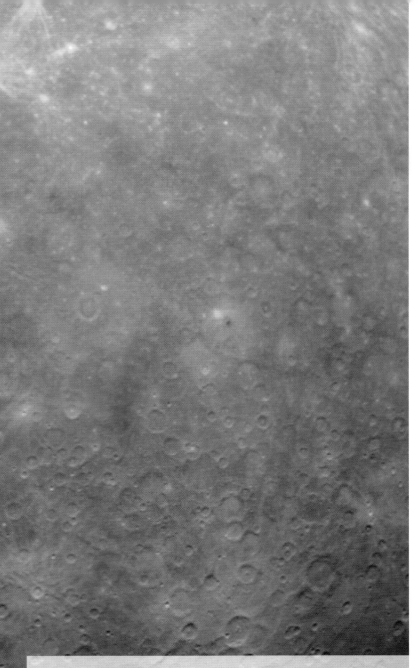

In April 1972, Apollo 16 astronaut Charles M. Duke, Jr., collects impact debris near the rim of the moon's Plum Crater.

It's All in a Name Eros asteroid, Copernicus crater on the moon, Maat Mons on Venus—where do all these names come from? The International Astronomical Union has been in charge of planetary nomenclature since 1919. As features are discovered on a planet, members of the naming groups composed of scientists approve names, ensuring that set rules for each planet are followed. On the moon (large image), large craters are named after deceased scientists, artists, and scholars; small-impact craters are named after villages of the world with populations less than 100,000. On Venus, all features are named for women; impact craters are named for deceased famous women. In order to prevent controversy, features are not named after living individuals.

asteroid or comet collisions, and the new insights into the cratering process enabled geologists to turn up dozens of impact structures on our home planet. We now know of more than 170 such impact craters pocking the surface of the globe.

Mapping the moon's craters and comparing their distribution across the lunar surface to the ages of the rocks sampled by the Apollo astronauts enabled geologists to determine the relative timing of impact episodes. The age-dated craters revealed that the inner solar system—Mercury, Venus, Earth, moon, Mars, and even Jupiter's icy satellites—was pummeled by an avalanche of colossal impacts through the first half billion years of its history. This bombardment, ending about 3.9 billion years ago, left these planets and moons heavily scarred. For example, on the moon's southern highlands and most of its far side, craters overlap craters so thickly that the original crust has been largely obliterated. The biggest of these crater-forming events formed multi-ringed basins more than a thousand miles across and flooded them with lava released by these deep wounds in the lunar crust on Earth. Most of the terrestrial evidence of that heavy bombardment has been erased by erosion and active plate motion.

Making the Moon

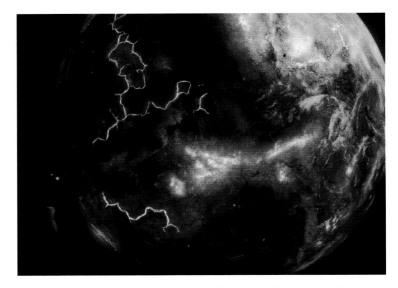

THE MOON'S VERY EXISTENCE is probably due to the collision of a Mars-size planetoid careening into a young Earth more than four billion years ago. The titanic explosion pulverized and melted our planet's outer crust and catapulted a hot plume of vapor and molten rock, torn from deep within Earth's mantle, high into space. The cooling debris eventually formed a temporary ring girdling Earth, which gradually coalesced into the early moon. Chemical analysis of the lunar rocks returned by Apollo astronauts supports this theory of lunar formation. Even as the moon formed, it continued to be bombarded, as did Earth, with a rain of other leftovers—asteroids and comets—from our solar system's formation.

Once the moon's crust cooled and solidified, its rocks began to absorb the history of this bombardment, like a kind of solid-state tape recorder. We can play back that record today, the lunar craters giving mute testimony to our celestial neighborhood's distant and violent past.

NASA's planetary spacecraft have found evidence of this early bombardment on the surfaces of all of the planets and moons of the inner solar system, telling us how many asteroids and comets circulated through the solar system early in its history. On younger portions of the moon's surface, the smaller, scattered craters we see match very well with the actual traces of impacts. The solar system's craters enable us to deduce how many hits the Earth should have sustained in the last few million years and to compare those estimates with the actual signs of impacts around the globe.

After forming from the orbiting debris ejected by the giant impact, the moon differentiated into layers—heavy materials sank to the interior to form the core and the mantle, while the less dense materials formed the outer lunar crust. We have many samples of the lunar crust returned by the Apollo missions, but we have limited knowledge of the chemistry of the rocks that make up the lunar interior. These rocks hold clues to the chemical makeup of the body that struck Earth, and may be found in the moon's South Pole–Aitken impact basin, a target for future lunar explorers.

This artist's rendering shows a Mars-size body hitting Earth (right) and the cooling and formation of the moon (above). The heat produced by the impact and the moon's formation left its surface covered in an ocean of magma (below). As the surface cooled, impacts began to leave scars on its surface, many flooding with lava to form the lunar maria, or seas.

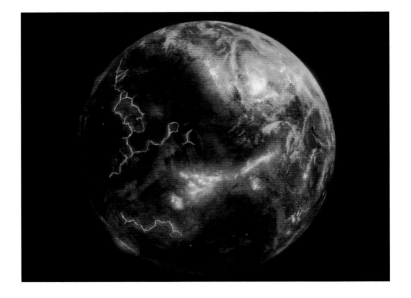

Return to the Moon NASA has ambitious plans to return to the moon for more scientific study, using both manned and unmanned missions. Although Apollo data changed our ideas on the formation of the moon, many questions remain, such as determining in more detail the impact history of the moon. This is very important for understanding Earth—the impact bombardment Earth underwent early in its history has left no visible trace but likely did affect how Earth evolved. By age dating many of the larger lunar craters, we can determine how frequently impacts occurred in early solar system history and use this to calibrate our understanding of impact episodes throughout the solar system.

Life in the Asteroid Swarm: *Living With Disaster*

W E LIVE TODAY ON A PLANET that orbits amid an asteroid swarm. These asteroids, which can measure 40 yards or more across, or larger, can penetrate the atmosphere to produce an explosion and a crater. Roughly once every 500 to 1,000 years, an object strikes Earth with the force of a megaton or more of TNT, like that which struck Tunguska, Siberia, in 1908. That stony asteroid, probably 50 yards across, exploded in the atmosphere in a one-megaton fireball, equivalent to about 60 Hiroshima atomic bombs. The blast and heat pulse flattened the forest over more than 800 square miles, igniting small fires in the largely uninhabited region.

NASA estimates there are about a million Near-Earth objects greater than 50 yards across—big enough to penetrate our atmosphere—but the bulk of the risk comes from those objects larger than 0.62 miles in diameter; NASA estimates there are about 1,100 of those. None we have detected are headed for an imminent collision. Because space is so big and Earth is a small target, your lifetime risk of dying from an asteroid impact is only about one in 40,000, about the same odds that you'll someday be killed in a plane crash. Astronomers are continually working to reduce that risk by extending our search programs. We can find asteroids and estimate their danger to us, but we don't have a plan for diverting one. Deciding what to do will be a global challenge, and no one knows how much time we'll have to take action.

Near-Earth asteroid Itokawa (above), a third of a mile long, was explored by the Japanese Hayabusa spacecraft in 2005. Itokawa appears to be a "rubble pile" asteroid, reassembled out of large and small framents after a catastrophic collision with another object. Its surface gravity is just one ten-millionth of Earth's.

NEAR-EARTH ASTEROIDS			
Asteroid	Possible Impact Date	Size	Probability of Striking Earth
99942 Apophis	2036	820 feet	22 in 1 million
2008 AP4	2089-2100	1,279 feet	40 in 1 million
1994 WR12	2054-2102	131 feet	100 in 1 million
2007 VK184	2048-2057	426 feet	340 in 1 million
2000 SG344	2068-2101	360 feet	1,800 in 1 million

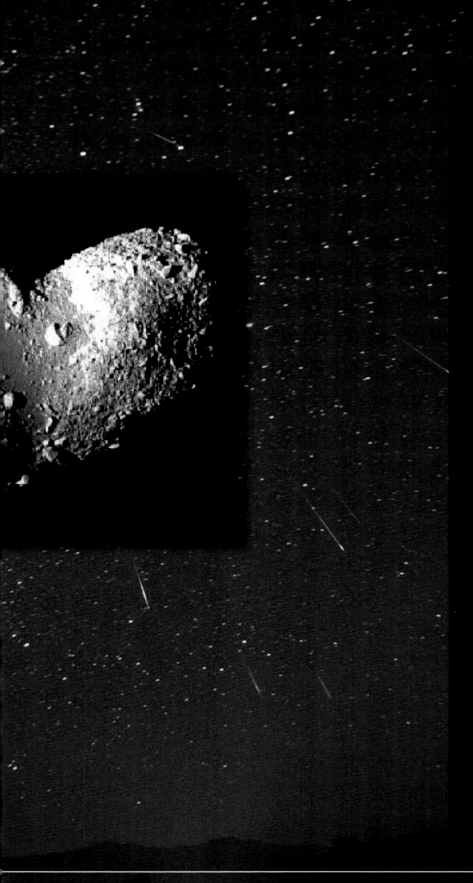

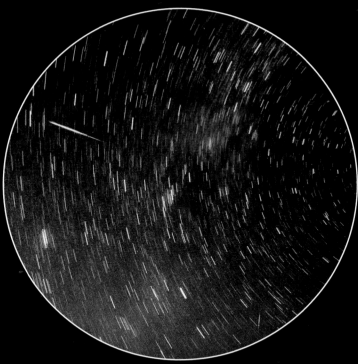

WHAT IS A METEOR SHOWER?

O ur Earth sweeps up about 55 to 110 tons of cosmic dust every day, most consisting of tiny particles, called micrometeoroids. No bigger than the period at the end of this sentence, they result from asteroid collisions or are carried from comets as the sun warms their icy, dusty surfaces. When these sand-grain-size, or smaller, particles hit Earth's atmosphere at speeds ranging from 7 to 40 miles a second, friction with air molecules instantly heats them white hot. Most are vaporized at altitudes of 50 miles or so; we see them as "shooting stars." Several times each year, Earth passes through the dust trail left in the wake of a comet, lighting up the sky with brighter, more numerous meteors. Such a perennial display, averaging 60 to 100 or more meteors per hour, is called a meteor shower. Showers are named for the constellation in the sky from which the dusty meteors appear to emanate; examples include the Perseids, Leonids, and Geminids. Visible as they circle nighttime Earth, the meteors burn up in the darkened atmosphere of our planet.

Chain Reaction: *Faraway Crash Landing*

SEE ALSO: *Deep Impact* 32

WE DON'T HAVE TO IMAGINE what a cosmic collision looks like. In 1994 our species had a ringside seat for the dramatic impact of comet Shoemaker-Levy 9 (SL-9) on Jupiter. Comet SL-9 was discovered in orbit around Jupiter in March 1993 by astronomers Eugene and Carolyn Shoemaker and David Levy. The comet had passed very close to Jupiter the year before, and the gas giant's gravity had torn the comet into at least 20 fragments, like so many cosmic pearls on a string. From July 16 to 22, 1994, these remnants, some as wide as 1.2 miles in diameter, crashed into Jupiter's banded upper atmosphere. The first fragment struck with the energy of 225,000 megatons of TNT; the impact plume rose more than 600 miles above the clouds. Additional impacts produced immense fireballs and vapor plumes, leaving behind Earth-size soot rings that marred the cloud tops for weeks. Linear crater chains seen on the Jovian satellites Callisto and Ganymede provide graphic evidence that similar comet impacts are commonplace throughout the solar system. Our robotic exploration of the planets has taught us more effectively than any earthbound evidence that we live in a cosmic shooting gallery.

The disrupted comet Shoemaker-Levy 9 (right) was imaged by the Hubble Space Telescope in 1993, showing the "string of pearls" appearance of the fragments, each just a few miles across. The fragments later struck Jupiter's atmosphere and dramatically showed the power of the impact process that has shaped the planets.

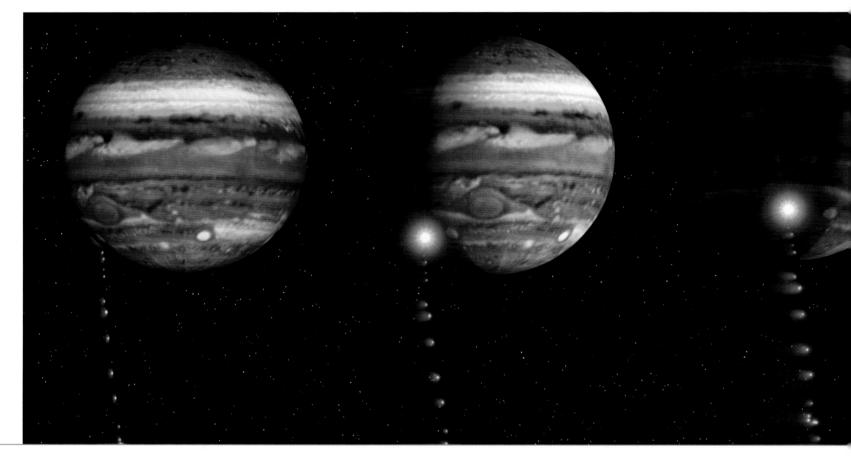

Ringside for a Collision This series of digital paintings depicts the titanic collision of comet Shoemaker-Levy 9 fragments with Jupiter's southern hemisphere. From Earth, the initial impact occurred behind the planet's visible horizon (far left). Instruments detected the flash of the image (center images) on Jupiter's night side. The white-hot fireball ejected a plume of hot, glowing gas nearly 2,000 miles above the cloud tops. The 18,000-degree fireballs and cooler plumes lasted just minutes before collapsing back into Jupiter's cloud bands, where comet debris spread outward like ripples in a pond (far right).

Repeat Performers Preserving the evidence of multiple impacts, the 10.5-mile-wide Aorounga crater in Chad (large image) was formed by the impact of an asteorid or comet several hundred million years ago, to the left of center in this shuttle radar image; two other fainter circular structures delineated by troughs are seen in the center and right of center of the image. Similar chains of craters can be seen on other planets, including the 13-crater chain on Jupiter's moon Ganymede (inset). Chains of craters form when the impactor breaks up shortly before impact.

Catastrophic Impact: *The Chicxulub Debate*

PERHAPS THE MOST FATEFUL, if not the largest, crater on Earth is Chicxulub, a 110-mile-wide multiringed basin located on the northwestern tip of Mexico's Yucatán Peninsula. Chicxulub is today invisible at the surface, but its formation shaped the very course of life on Earth. In 1980, geologists Luis and Walter Alvarez and their team found an elevated amount of the scarce element iridium in a sediment layer dating back 65 million years. This thin clay layer marked the end of the Cretaceous and the beginning of the Tertiary period in Earth's geologic history, a demarcation referred to as the "K/T boundary." Iridium, very rare in Earth's crust, is much more plentiful in meteorites —and asteroids. Finding enhanced iridium in this K/T rock layer at more than 80 locales worldwide led to the Alvarezes' conclusion that the deposit was formed by the global effects of an asteroid or comet impact.

Geologists had noted the jumbled beds of 65-million-year-old sediments surrounding the Gulf of Mexico, along with matching deposits of once molten impact glass (tektites) in Haiti and began

zeroing in on the K/T crater. They found it by recognizing ring-like arcs in gravity surveys of the Yucatán subsurface, and confirmed an impact at Chicxulub by examining rocks brought up in exploratory drill cores by the Mexican national oil company. The crater is the closest thing we have to a smoking gun, the cause of one of the most dramatic mass extinction events in Earth's history.

Chicxulub is buried under about half a mile of marine limestones laid down since the impact, but the story it tells is unmistakable—and terrible. Its central depression, buried rim, and outer rings match those of impact features on the moon, Mercury, and Mars; what happened in the Yucatán 65 million years ago happened thousands of times in solar system history. A comet or asteroid some six miles across struck the shallow sea there at cosmic velocity, creating giant tsunamis that washed wave deposits dozens of miles inland. An intensely hot fireball scorched the region, killing everything within hundreds of miles of ground zero, and threw millions of tons of dust into the atmosphere. Global darkness killed vegetation. Water vapor and carbon dioxide ejected by the impact heated the Earth by nearly 20°F, and a massive acid rain poisoned the upper oceans. The world's dead forests burned, and soot spread across the globe. Roughly 80 percent of the planet's living species disappeared in the worldwide catastrophe. Among the few survivors were the earliest mammals—our ancestors.

Our planet's scars are a reminder that the potential for a disastrous impact still exists today (but one that we, with improving space technology, may be able to prevent). Less evident from orbit, but still visible to a careful astronaut observer, are signs of outbursts from within Earth that sometimes rival the destructive equal of cosmic events like Chicxulub.

The dinosaurs (*Masiakasaurus*, left), Earth's dominant life-form for 185 million years, were probably doomed by the catastrophic ecological effects of the Chicxulub impact 65 million years ago: Global fires, choking dust, and eventual worldwide cooling.

Hidden Impact The now buried Chicxulub impact crater lies beneath 1,000 to 3,000 feet of limestone deposits that make up the Yucatán Peninsula (large image) and the adjacent seafloor. The initial multiringed impact basin was soon eroded and blanketed in marine sediments and lost from view. Curved fractures in the limestone have helped form freshwater springs and sinkholes, called cenotes, that today mark the buried crater rim. Long hidden underneath tropical savanna vegetation in northern Brazil, lies the evidence of another ancient collision (inset). Roughly 220 million years ago, geologists estimate, a meteorite struck, forming the Serra da Cangalha crater.

Lava Floods: *Sheets of Fire*

SEE ALSO: *Scorching Valleys* 120

MASSIVE LAVA FLOWS have periodically covered huge areas of Earth with successive thin sheets of molten rock. In the Pacific Northwest, remnants of one such flow, the Columbia River flood basalts, cover an area of 63,300 square miles. Layer after layer of swiftly erupted lavas piled up to a depth of more than two miles! The total volume of this Columbia River basalt group is 42,000 cubic miles; the Grande Ronde flow, 85 percent of that total, erupted enough molten rock in less than a million years to bury the continental United States in 40 feet of dense, black lava. No volcanic episode in recorded human history remotely approaches this fantastic rate of eruption.

The Columbia basalts, called flood basalts (after the type of fluid, upper-mantle melts involved), were deposited about 16 million years ago, with the last activity ending about 6 million years ago. These flows resemble the dark basalts that filled the lunar maria, or seas, billions of years ago. Flood basalts are, in fact, visible on all the rocky planets and moons, including Jupiter's moon Io. Other flood basalt regions on Earth

The Columbia River Gorge in Washington State slices through the thick layers of the Columbia flood basalts, laid down over ten million years. Slowly cooling basalt crystallized into tall, hexagonal columns of lava.

Basalt Floods Basalt is a black volcanic rock that is found on the surfaces of Earth (seen here at Kilauea volcano in Hawaii), Mars, Venus, the moon, likely Mercury, and even the asteroid Vesta. It is rich in iron and magnesium. Basalts can vary in their chemical composition, depending on the chemistry of the magma it came from, which in turn depends on the chemistry of the rocks from the interior of the planet that melted to form the magma. When a basaltic lava erupts, it has many different forms, depending on whether it erupted under water or on a surface, how quickly it erupted, and the exact chemistry and amount of water in the magma. For example, the very thick Columbia River basalt flows formed columnar shapes as they cooled.

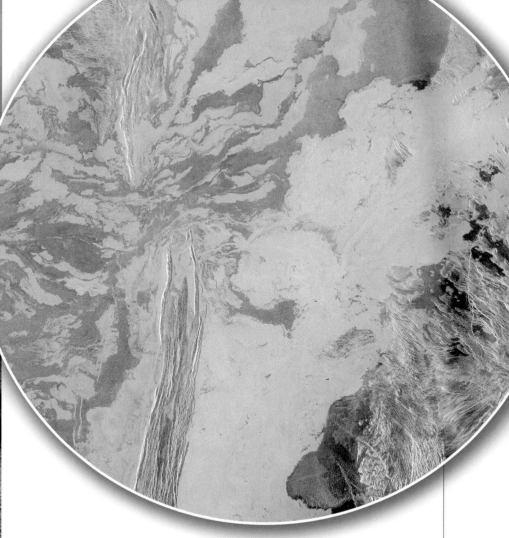

Venus's Lada region (above), located in its southern hemisphere, preserves a large lava flow field similar to the Columbia River basalts. Both were laid down by eruptions on a much larger scale than the Kilauea flow pictured in the large image at left.

include the Laki flow in Iceland, the Deccan Traps in India, and the Siberian Traps near Norilsk.

Flood basalts are the largest volcanic events in Earth's history; by comparison, Mount St. Helens in 1980 erupted a third of a cubic mile of ash and lava. The Deccan flood basalts, dated to between 60 and 65 million years ago, may have also contributed, through the huge outpouring of volcanic gases and subsequent climate change, to the demise of the dinosaurs. And the Siberian Traps eruption, 251 million years ago, is the prime suspect in the biggest extinction event in Earth's history, the demise of 90 percent of all living species at the end of the Permian period.

LAVAS

As a planetary scientist, I spend most of my time looking at images of far-away planetary surfaces, trying to make sense of how a volcano formed or unravel the geologic history of a complex surface. So one of the most enjoyable parts of my job is to be able to get out and actually look at a volcano up close. I have done fieldwork on lava flows at Mount Etna in Sicily, Kilauea volcano in Hawaii, in Iceland, and on volcanoes in California and Oregon. Places like the Laki fissure in Iceland (main image), which produced major eruptions in 943 and in 1783-84, are important for understanding volcanic eruptions, but also beautiful. Most of my fieldwork has been on lava flows that erupted tens, hundreds, or even thousands of years ago, but I have also done work on actively erupting flow fields in Hawaii. Active flows are exciting—the heat is amazing! We study lava flows in the field to better understand how they form. As a lava flow moves away from the vent from which it is erupting, the outer surface, or the crust, cools over the still moving, liquid interior of the flow. Features that are preserved on this surface crust, including channels, changes in texture, and bumps called tumuli, can provide a lot of clues as to how the liquid lava was moving under the cooling crust of a flow, including information on how fast the lava was erupting and moving downslope. Large lava flow fields on Earth are old and heavily eroded, and there is still much scientific debate over how quickly the lava flows were erupted. By looking at younger flows where we have better preserved surfaces, by watching how a flow forms in real time, and by comparing them to the uneroded flood basalts on the surfaces of other planets, we are trying to understand how these massive eruptions occurred on Earth.   —Ellen Stofan

Dangerous Job The large rock my daughter Sarah and I are holding is a volcanic bomb—a rock thrown out of the top of Mount Etna during an eruption that occurred a few days before our visit. This time, I was there as a tourist showing off my favorite volcano to my family—but we got to see freshly steaming lava flows that had erupted onto the snowy top of Mount Etna. A rock like this can cause injury to anyone too close to the summit of a volcano during an eruption.

Supervolcanoes

ONE OTHER CATACLYSMIC, but fortunately very rare, type of volcanic outburst is the supervolcano, a popular term for a single, enormous eruption that dwarfs anything in human experience. The evidence for such super-eruptions consists of enormous igneous deposits covering thousands of square miles in Japan, New Zealand, the western United States, and Indonesia, the result of an ash fall hundreds of feet thick. California's Long Valley Caldera, for example, erupted at least 140 cubic miles of magma about 700,000 years ago. More than 12 cubic miles of that total was widely dispersed as ash. From the caldera in the central Sierra Nevada, the ash (called the Bishop tuff) fell as far away as eastern Kansas and Nebraska, accumulating in drifts up to 30 feet thick.

Supervolcanoes have also been found on Mars and Jupiter's moon Io in our solar system. Understanding these volcanoes may give us an idea of the danger we face from terrestrial volcanic centers like Yellowstone and Lake Toba in Indonesia, areas that lie atop hot spots—places in Earth's mantle where magma wells up—that will undoubtedly erupt again.

We will need every bit of human adaptability and ingenuity for our civilization to survive such a catastrophic event.

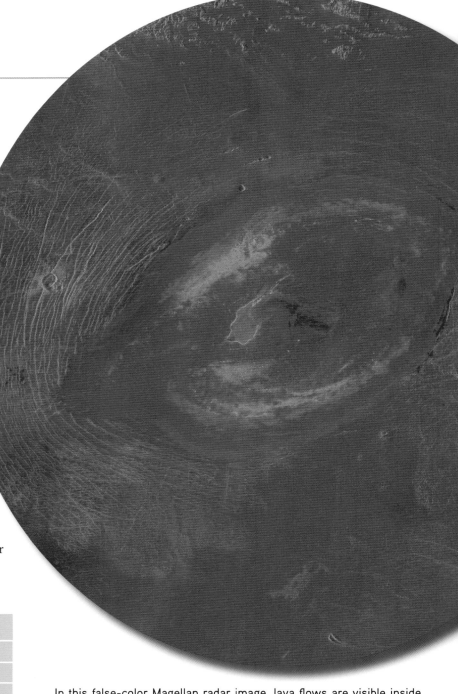

In this false-color Magellan radar image, lava flows are visible inside and surrounding Venus's mega-caldera Sacajawea Patera, one of the largest volcanoes in the solar system. Its volcanic depression, or caldera, is 252 miles across and more than a half mile deep, 200 times larger than Earth's largest active volcano.

FLOOD BASALTS

Location	Area (square miles)
Deccan Traps, India	200,772
Paraná Basin, Brazil	463,320
Karoo, South Africa	54,054
Siberian Traps	579,150
Columbia River, Washington State, U.S.A.	63,320

A Sleeping Giant Yellowstone's geyser fields show this vast supervolcano, which has erupted repeatedly over the past two million years, is still active. More than 20 major eruptions occurred between 70,000 and 160,000 years ago. But these were dwarfed by an earlier outpouring of 240 cubic miles of material from Yellowstone's magma chamber. That cataclysmic eruption 640,000 years ago blanketed most of the central U.S. with ash. Today, geologists carefully monitor its activity.

Averting a Cosmic Catastrophe

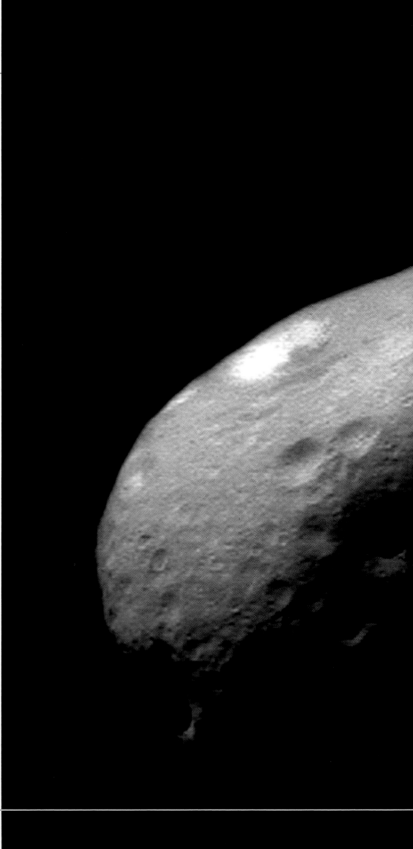

A STEROIDS AND COMETS have battered Earth thousands of times during its 4.6-billion-year history. We will be struck again. If an asteroid larger than half a mile across struck the Earth, the explosion would throw enough dust into the atmosphere to shut down agriculture for a year or more, possibly collapsing civilization. NASA is conducting a survey called Spaceguard to find 90 percent of all near-Earth objects (NEOs) larger than 0.62 miles across by 2008.

That survey has found about 800 such objects so far out of the roughly 1,100 that are estimated to exist. None are on a collision course. The searchers operate telescopes in Hawaii, Arizona, New Mexico, and California, each scanning the background of stars for moving pinpoints of light that might represent a new asteroid or comet. Scientists then calculate an orbit to assess any risk.

Many NEOs smaller than these civilization-busters will penetrate the atmosphere; the size threshold is about 160 feet across. In 1908, for example, an object lit up the atmosphere over Tunguska, Russia, and exploded high above unpopulated Siberia. The blast flattened nearly 800 square miles of forest. Such an impact could destroy a city. Congress has directed NASA to conduct a more sensitive search by 2020 to find 90 percent of all NEOs larger than 450 feet across. We know of only a few thousand of the roughly 100,000 such objects. If we keep looking, and do some prudent planning, we'll make sure a "big one" never threatens Earth again.

Radar telescopes like NASA's 228-foot Goldstone antenna (at left) furnish precise orbital tracking of near-Earth asteroids and can determine their shape and size. Goldstone and the larger Arecibo radar in Puerto Rico are key facilities in our search for potential impactors.

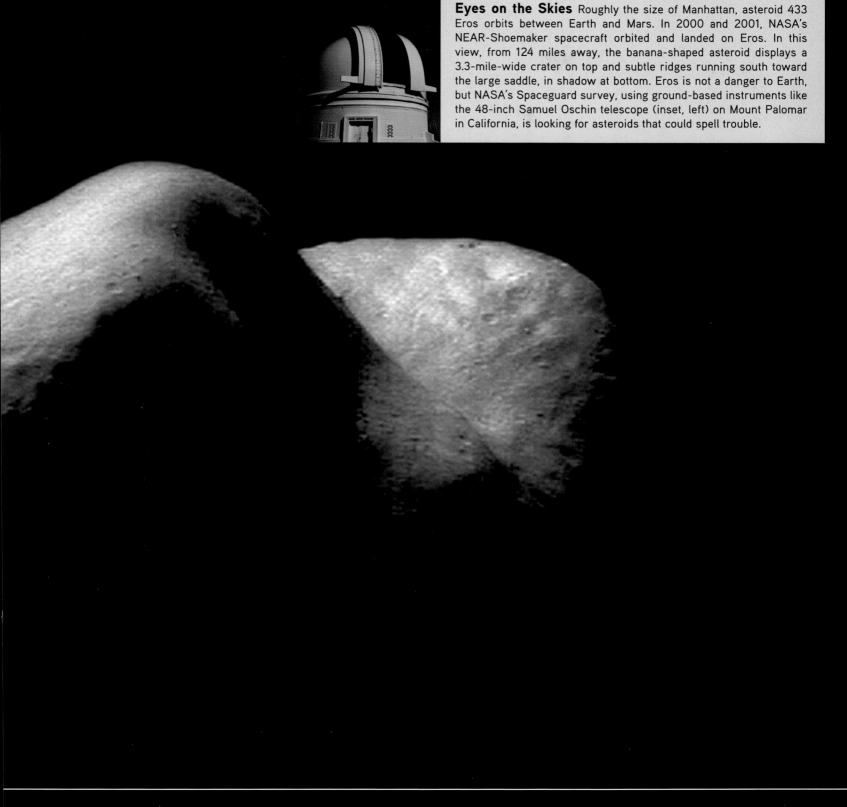

Eyes on the Skies Roughly the size of Manhattan, asteroid 433 Eros orbits between Earth and Mars. In 2000 and 2001, NASA's NEAR-Shoemaker spacecraft orbited and landed on Eros. In this view, from 124 miles away, the banana-shaped asteroid displays a 3.3-mile-wide crater on top and subtle ridges running south toward the large saddle, in shadow at bottom. Eros is not a danger to Earth, but NASA's Spaceguard survey, using ground-based instruments like the 48-inch Samuel Oschin telescope (inset, left) on Mount Palomar in California, is looking for asteroids that could spell trouble.

SURFACES IN MOTION

The destructive motion of an earthquake can topple houses and rip up roads, as evident in this image of earthquake damage after a quake in Anchorage, Alaska.

Continents, Crusts, & Collisions

SEE ALSO: *Mountain Formation* 73

FOR BILLIONS OF YEARS, change on Earth's surface has been driven by a process called plate tectonics. Several hundred million years ago, all of the land on Earth was combined into one large supercontinent called Pangaea, surrounded by a huge ocean. Over roughly the last 230 million years, the pieces of this landmass separated and drifted apart, riding on convection currents in the interior, and producing the current, familiar pattern of continents scattered around the globe. The fact that the edges of some continents look like puzzle pieces that could fit together led scientists to first suggest the theory of continental drift, which was refined into the theory of plate tectonics during the 1960s.

Of course, the pattern of continents we see today is not stable—they are still in motion. The manifestations of this motion are active faults, seams where most of the motion—and Earth's geologic activity—is concentrated. Faults and lines of volcanoes trace the edges of the plates, and the motion of the plates causes earthquakes, great and small, which occur daily around the globe.

Driving this motion is heat within Earth, which is produced by the decay of radioactive elements like potassium, uranium, and thorium. This decay produces density differences in the rocks of the mantle, with hotter rocks being lighter. Warm rock can actually behave like plastic—it can flow, although very slowly. These density differences drive convection in the interior, where hot, less dense material from near the core is constantly rising toward the surface, then cooling and sinking back. This large-scale process of convection—moving plumes of hot rock—is thought to drive plate tectonics.

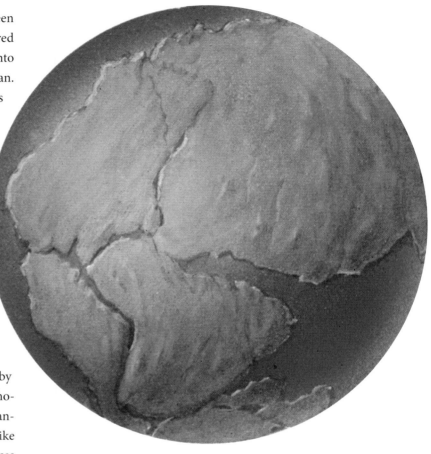

The arrangement of the continents on Earth's surface is always changing—inch by inch. The breakup of the supercontinent of Pangaea (above) is shown in three stages at right, with the final globe on the far right illustrating a future breakup of Africa along the East African Rift.

---

2 BILLION YEARS AGO
Enough oxygen accumulates in the atmosphere, probably through volcanic activity, for life-forms to evolve and flourish.

225 MILLION YEARS AGO
The collision of the major continental plates forms a supercontinent, known as Pangaea.

65 MILLION YEARS AGO
A meteorite smashes into Earth creating the Chicxulub crater and leads to the extinction of the dinosaurs.

50 MILLION YEARS AGO
The Indian plate, originally situated about 4,000 miles south in the Pacific Ocean, collides with the Eurasian plate and forms the Himalaya.

Earth's plates comprise the rigid lithosphere—the shell-like layer that rides on the asthenosphere, which is at least partially melted and can flow.

Earth's surface is constantly in motion, and its plates interact in three ways. Where they spread apart, new crust forms from upwelling of molten rock, as at the undersea mid-ocean ridges. Where plates collide, they cause one plate to descend beneath the other back into Earth's interior in a process called subduction. Subduction occurs at many continental boundaries, as at Japan where the Pacific plate dives beneath the Eurasian plate. Subduction constantly recycles Earth's older crust, and the forces associated with the process are thought to be the most significant in keeping plate motion going. Plates can also interact by simply sliding past each other, which produces large faults and dangerous earthquake zones, such as the San Andreas in California at the boundary between the North American and Pacific plates. When two continents run into each other, the continental crust is generally too light to subduct, so the two plates crumple into each other forming mountain belts, like the massive Himalaya at the Indian-Asian plate collision and the Alps at the intersection of the African and Eurasian plates.

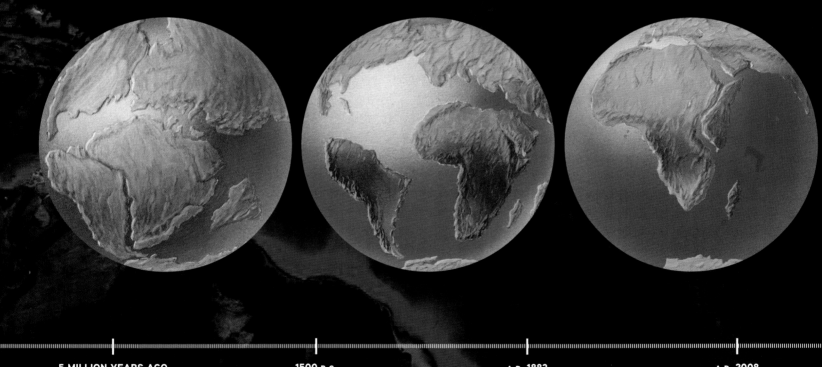

5 MILLION YEARS AGO
The Colorado River begins cutting down through the uplifting Colorado Plateau to form the Grand Canyon.

1500 B.C.
The ancient Minoan trading island of Thera, now called Santorini, erupts. Ash and pumice from the eruption have been found in Israel and Egypt.

A.D. 1883
The Indonesian volcano Krakatau erupts. The mountain spews ash 10 miles into the air and can be heard 2,200 miles away in Australia.

A.D. 2008
The Sichuan Province of China is struck by a massive earthquake that kills tens of thousands and destroys vast amounts of land.

Finding Faults Extending from Baja California more than 800 miles north, the San Andreas Fault, as seen in this topographic radar image (inset) from the space shuttle, forms the boundary between the North American plate on the right, and the Pacific plate on the left. Movement along the fault causes earthquakes like the famous 1906 San Francisco quake. While Earth is the only body in the solar system known to have an active system of plate tectonics, Enceladus (large image) exhibits surface cracks in its icy exterior that may help us better understand faults at the edges of plates on Earth.

The signs of plate tectonics are linear mountain belts, chains of volcanoes, and long, strike-slip faults, organized into patterns that define plate boundaries. Earthquakes are also concentrated at plate boundaries, but we have yet to deploy a seismic network on a planetary surface (other than on the moon) that would allow us to map earthquake epicenters. We can also identify tectonics by mapping the patterns of magnetization of the crust.

Surprisingly, scientists have been unable to identify plate tectonics on any other body in the solar system, although many planets and moons exhibit evidence of surface movement and feature significant mountain ranges. We cannot rule out the possibility that other bodies, particularly Mars and Venus, may have been shaped by plate tectonics in their early histories, with the evidence now obscured by later geologic events. Yet when it comes to studying the process that really controls how our own planet deforms—plate tectonics—we find no easy parallels. Earth appears to be the only world with an active plate system.

What is it about Earth that has allowed its surface to break into numerous plates? Part of the answer lies in its size and composition. Earth is the largest terrestrial (rocky) planet, and thus has the most internal heat, which in turn drives interior convection. It also has abundant surface water, which aids subduction. Venus is large like Earth, but it is very dry. Mars, like the bodies of the outer solar system, is small and thus retains less warmth from heat-producing elements. But early in its history, Mars would have had higher levels of interior heat. The Mars Observer spacecraft carried a magnetometer that detected odd patterns of magnetized crust on the planet's surface, leading some scientists to speculate that Mars was affected by plate tectonics early in its history. This raises the question of how plate tectonics started on Earth, and when and how it may end. Studying the other bodies of the solar system in detail may help to address these questions.

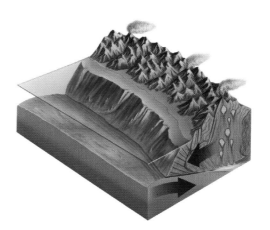

Subduction zones occur when one plate is thrust beneath another, producing a deep trench on one side and a volcanic arc on the other.

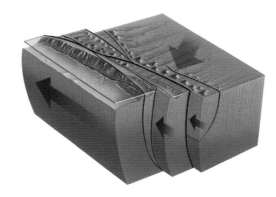

At transform boundaries, two plates slide past each other producing damaging earthquakes, but neither producing nor destroying crust.

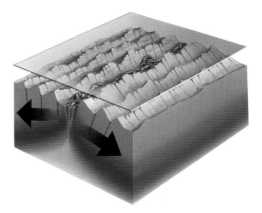

Spreading centers are linear zones separating two plates, along which new crust is created by rising magma, which spreads outward.

Dynamic Planets

SEE ALSO: *Ridges* 70

F PLATE TECTONICS does not produce faults and mountain ranges on other planets and satellites, then what does? Even planets without moving plates can have interior motion that can cause crustal motion. On smaller moons, crustal stresses can be produced by gravitational tides caused by the moon's motion around the parent planet—such as on the icy satellites of Jupiter and Saturn. As the satellite orbits the planet, the planet tugs gravitationally on the satellite—more when it is closer and less when it is farther away. On Earth, this same sort of gravitational tug (caused by the moon) produces ocean tides. Around Jupiter and Saturn, the pull of the large planet on the small satellite is enough to flex the interior of the satellite, causing it to heat up and produce geologic activity on its surface.

The basic types of tectonic features that we see on the planets are not unlike those that we observe on Earth. We have observed features produced by extension (pulling apart), compression (coming together), or shear (moving past each other). The overall shape and dimensions of the resulting tectonic features can help us to model how a surface—on any planet—deforms.

One of the solar system's most intriguing rifts is Valles Marineris on Mars, a vast system of canyons that is roughly the length of the United States and more than four times as deep as the Grand Canyon. Valles Marineris started out as grabens, or rifts, that were enlarged over time by erosion and collapse. The complex geography of the region is characterized by multiple, interlocking canyons. The cause of the extension is probably related to the Tharsis bulge—a huge pile of volcanic material that is located on the opposite side of Mars. This immense stack of lava produced stresses in the crust of Mars, causing it to break.

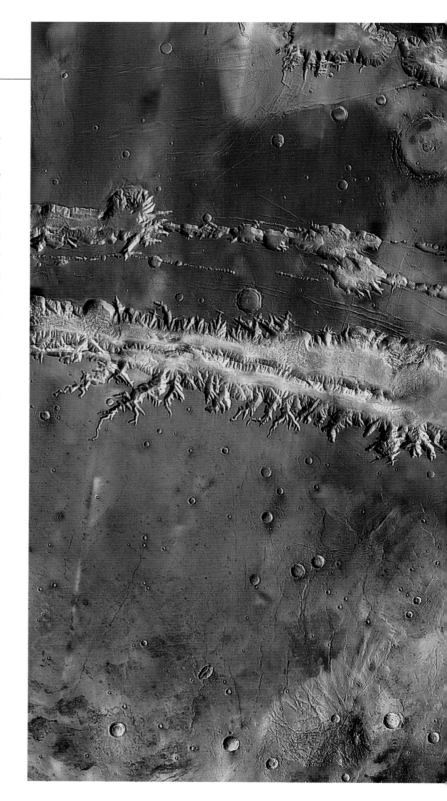

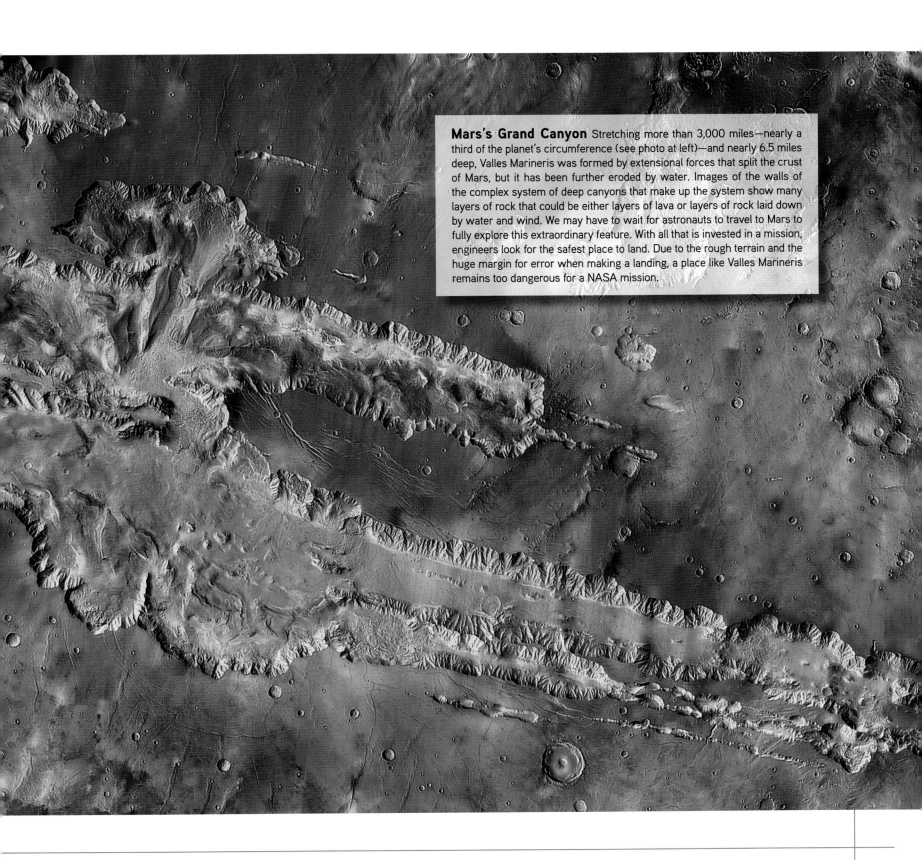

Mars's Grand Canyon Stretching more than 3,000 miles—nearly a third of the planet's circumference (see photo at left)—and nearly 6.5 miles deep, Valles Marineris was formed by extensional forces that split the crust of Mars, but it has been further eroded by water. Images of the walls of the complex system of deep canyons that make up the system show many layers of rock that could be either layers of lava or layers of rock laid down by water and wind. We may have to wait for astronauts to travel to Mars to fully explore this extraordinary feature. With all that is invested in a mission, engineers look for the safest place to land. Due to the rough terrain and the huge margin for error when making a landing, a place like Valles Marineris remains too dangerous for a NASA mission.

A Rift in Earth

WHEN THE SURFACE OF A PLANET stretches, the crust literally breaks apart, producing cracks called normal-faults; blocks of crust drop down to form flat-floored valleys, called grabens or rifts. On Earth, the Basin and Range province in Nevada, which has been extending for millions of years, provides an example of the crust pulling apart over a broad region, producing a series of normal-fault-bounded ridges and down-dropped valleys. On a larger scale, the African plate is also trying to break apart, producing the more than 3,000-mile-long East African Rift System that extends from Lebanon to Mozambique. The rift is very active geologically, and if it continues to spread, it will eventually create a new ocean and a new island.

In both of these cases, the crust extends by both flowing (in the lower, warmer portion of the crust) and faulting. Only the top 6-9 miles of the crust is cold and rigid enough to break and form faults; below that, the crust is warmer and more plastic and gives way by flowing. Movement along faults can occur in a slow, gradual fashion, called creep, but eventually enough strain builds up so that a more sudden break occurs, resulting in an earthquake. How devastating the earthquake will be is governed by the length of the fault section that breaks, as well as how deeply the break goes. In the cases of both Nevada's Basin and Range and East Africa, the extensions are accompanied by upward-welling heat, causing melting of rock beneath and within the extending crust. Magma, rising through the breaks in the crust, makes its way to the surface, creating a series of volcanoes along the extension zone.

THE SOLAR SYSTEM'S GREAT RIFTS

Rift	Planet	Length (miles)
East African Rift	Earth	4,000
Parga Chasma	Venus	6,000
Valles Marineris	Mars	4,000
Hecate Chasma	Venus	4,800
Mid-Atlantic Rift	Earth	6,000

The Great Rift This triangular-shaped piece of land (at left) is the Sinai Peninsula, with the Gulf of Suez to its left and the Gulf of Aqaba to its right. The gulfs and the Dead Sea are the extensions of the same rift that resulted in the opening of the Red Sea. It splits the Arabian and African plates along a 3,600-mile rift valley extending from Lebanon to the Red Sea. The split began about 30 million years ago and continues, inch by inch, today.

Iceland's Thingvellir (above) is a natural amphitheater formed by extensional faults above the constantly spreading Mid-Atlantic Ridge. The Icelandic parliament has met here each year since A.D. 930.

Rifts are also common on Venus, Mars, and the icy satellites of the outer solar system. The surfaces of these bodies have undergone extension, though not sufficient to break their crusts into distinct plates. In addition to Valles Marineris, other evidence of limited extension on Mars includes groups of fractures and grabens, like the Cerberus Fossae near the Elysium volcano. While these are tectonic grabens formed by extension of the crust, they illustrate that volcanic activity can cause a surface to deform. The fractures of Cerberus formed when movement of magma associated with the Elysium volcano stretched the crust. High-resolution images of Cerberus grabens returned by the NASA Mars Odyssey and the NASA Mars Reconnaissance Orbiter spacecraft show how the younger extensional fractures cut across the plains of Mars, then were widened by subsequent collapses in the fracture walls.

Venusian Rifts

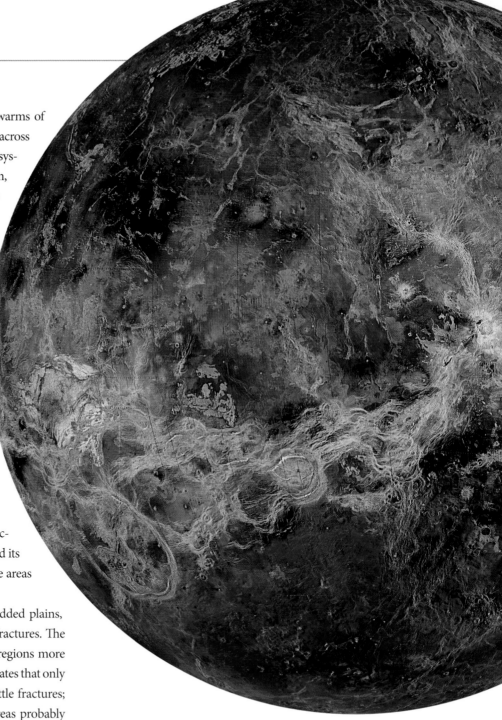

ON VENUS, EXTENSION HAS PRODUCED swarms of fractures, strange gridded plains, and rifts cutting across much of the equatorial region of the planet. The rift systems of Venus are among the most extensive in the solar system, forming families of faults and troughs that extend roughly the distance from New York City to Istanbul, Turkey. The rifts average about 155 miles in width, with depths of several miles. The Venusian rifts are more common and much longer than on Mars, indicating Venus has a much more active surface. But like Mars, the amount of extension on Venus is limited—new crust was not created, in contrast to the seafloor rifts on Earth. The Venusian rifts are probably related to interior convection, where the flow of mantle material transmits stress to the surface. However, on dry Venus, the lack of water in the rocks may prevent subduction—a key driver of plate tectonics—from getting started. The dry crust of Venus—too strong to subduct—may move around without breaking up into plates. The lack of water on Venus means that erosion is much reduced, preserving tectonic features there—and enabling lucky geologists here on Earth to analyze them in their pristine state.

Some areas of Venus are cut by extensive swarms of fractures. Some of these may have formed when hot magma forced its way into the crust, fracturing the surface, while others may be areas where the extension was not strong enough to create rifts.

One of the oddest tectonic regions on Venus is the gridded plains, low-lying regions cut by perpendicular sets of very narrow fractures. The fractures are spaced as little as a half mile apart, and cover regions more than 60 miles across. The narrow spacing of the fractures indicates that only a thin upper layer of the crust has been broken by these brittle fractures; the uniform nature of the gridded plains over such large areas probably indicates that they were originally covered by lava flows with very uniform mechanical properties, allowing this series of evenly spaced breaks.

In this topographically colored image of Venus's complex rift zones, yellows and reds indicate higher areas, and blues indicate troughs.

Troughs on Mars The long troughs of Cerberus Fossae on Mars (large image) expose layers of rock in the walls where the crust has pulled apart. Rocks from the walls tumble down the slopes to form the dark-toned debris of the talus deposits on the trough floor. Martian canyons form without plate tectonics, which is responsible for the fractures and troughs of Earth's Great Rift in East Africa (inset). Lakes, including Lake Edward, at right, mark much of the low-lying regions of the East African rift.

Icy Worlds

SEE ALSO: *Europa's Watery Secrets* 180

I N THE OUTER SOLAR SYSTEM, the satellites of the gas giants are made up of a combination of silica-rich rock and water ice. Farther from the sun, the satellites tend to be made of more ice and less rock. However, the extremely cold temperatures at these distances make for extremely strong water ice, allowing it to behave more or less like rock. Consequently, we find geologic features on the icy satellites that resemble tectonic features found on the rocky surfaces of Earth or Mars.

What causes these frigid surfaces to deform? The icy interiors of these outer planet satellites do contain radioactive elements, but not enough to produce internal heating and melting. What churns up the insides of these bodies, producing heat to drive geologic activity, is the presence of the nearby gas giant. A gigantic planet like Jupiter or Saturn tugs strongly at its satellites, producing tidal flexing or motion in the interior and creating frictional forces. Just as the frictional force of rubbing your hands together produces enough heat to warm your hands up—the tidally produced friction in the icy satellites is enough to cause the interior of the satellites to warm up enough to produce melting.

Images from the Voyager and Galileo spacecraft at Jupiter provided ample evidence that the surfaces of Europa and Ganymede, two of Jupiter's moons, have extended, forming belts of grooves that snake across the surface. When cracks form on these frozen bodies, a water-ice mush wells up through the voids, forming new crust like the ice on a New England pond. Multiple episodes of extension on both moons have created a crisscross network of fracture belts.

Jupiter's tidal forces, pulling on these moons repeatedly, creates at least partially melted interiors, whose motions extend and compress the surface. For example, Europa's surface rises and falls almost a hundred feet between high and low tides—placing a lot of stress on the surface. The fact that the older fractures do not align with the current tidal stress system led scientists to discover that the surface is rotating faster than the interior. The surface, we think, is detached from the interior by a liquid layer. This all-important subsurface liquid layer—an interior ocean—is of major interest to astrobiologists.

Voyager and Cassini documented similar grooves on the surfaces of some of the Saturnian satellites. Tiny Enceladus, only about 310 miles in diameter and orbiting inside the E ring of Saturn, has a surface resembling its larger Jovian cousin Europa. With heating provided by a combination of tidal stresses and its radioactive rocks, the surface of Enceladus is cut by fractures, some of them up to 124 miles long and more than a half mile deep. Cassini images of Enceladus have revealed the unexpected complexity of the fracture systems, as well as an intriguing region of fractures near the south pole. Some of these fractures near the center of the region appear to be very young and may even contain organic material.

Enceladus is not the only moon of Saturn with evidence of cracking and extension of its surface. The Voyager spacecraft imaged bright wispy regions near the southern pole of Dione (at left) that Cassini high-resolution images revealed as fractures with steep cliffs.

This Cassini image shows a 37-mile-wide crater cut by a fracture on the surface of Dione, a moon of Saturn.

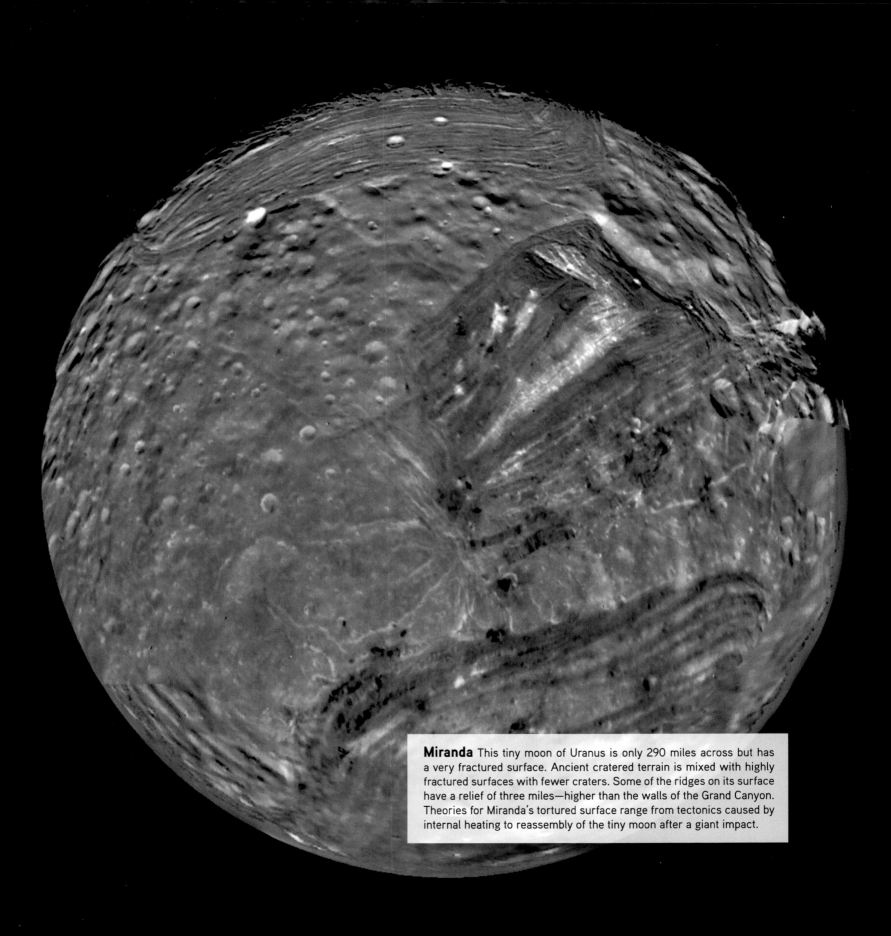

Miranda This tiny moon of Uranus is only 290 miles across but has a very fractured surface. Ancient cratered terrain is mixed with highly fractured surfaces with fewer craters. Some of the ridges on its surface have a relief of three miles—higher than the walls of the Grand Canyon. Theories for Miranda's tortured surface range from tectonics caused by internal heating to reassembly of the tiny moon after a giant impact.

Ridges: *Folds & Wrinkles*

COMPRESSIONAL STRESS OCCURS when the crust of a planet is pushed together, resulting in the formation of folds, and thrust or reverse faults. On the surface, these types of features typically look like ridges. Compressional ridges tend to be sinuous or S-shaped, whereas features formed by extension tend to be much straighter. Folds come in two types: anticlines, where the layers of rock bend upward and synclines, where the rock layers bend downward. How does something as hard as rock bend and fold? Under high temperature and pressure, rocks can deform plastically—change shape without breaking. If the rock is nearer the surface, colder, and under less pressure than deeper rock, it will tend to break into a fracture or fault. On Earth, this change in behavior occurs about nine miles below the surface. Folds in mountain belts on Earth are observed because the once deeply buried folds have been uplifted and eroded down, exposing the structure. Using planetary images, we can see only the surface, which in the case of Mars is sometimes eroded enough to allow us to see some internal rock layering; however, in the case of the icy satellites, Venus, the moon, and Mercury, the interior structures have to be guessed at from their surface shapes.

While most of the icy satellites record a history of extension, narrow compressional ridges have been identified on the Jovian moon Europa. In fact, it is surprising that more evidence of compression is not found on the icy satellites, given all the extension that is seen—to conserve area you have to have compression somewhere! But the folds found on Europa are very subtle and are likely produced by the tidal motions caused by Jupiter. Experts argue that these folds may be present on many icy satellites—but are hard to identify because they are low and broad (requiring just the right shading to see) and because they probably relax away to flatness relatively quickly in the ice-rich, plastic crust of most of the satellites.

On Mars, narrow, snakelike ridges cut the volcanic plains, looking like wrinkles in a blanket. These wrinkle ridges are also found on volcanic plains on the moon, Mercury, and Venus. Wrinkle ridges form due to shallow compression affecting a broad, relatively strong, thin portion of the crust.

While Mercury's surface mostly preserves a long history of volcanism and impacts, its surface is cut by a series of thrust faults, where one crustal section overrides another. These faults originated from compression of the crust that occurred very early in the history of Mercury. As discussed in the last chapter, the rocky planets formed from the impacts of many planetesimals very early in the history of the solar system, about 4.6 billion years ago. These impacts generated a tremendous amount of heat, resulting in an early phase where a planet is partially melted, followed by slow cooling. Most rocks contract as they cool, and the planets would have shrunk—decreased in radius—as they cooled. Many of the thrust faults on Mercury appear to have formed during its early history, when its crust would have contracted and compressed as it cooled.

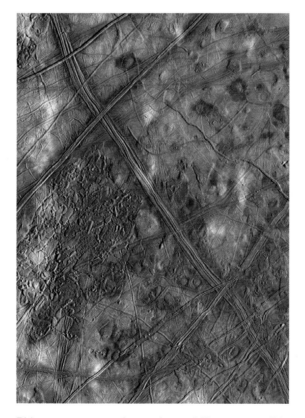

Ridges cut across the surface of Europa, possibly caused by motion of the subcrustal ocean. Some ridges and fractures on Europa are 1,850 miles long.

ON OTHER WORLDS

Compression in the Thaumasia region of Mars, a high volcanic plain located south of Valles Marineris, produced these long, winding wrinkle ridges (at right of large image) captured by the HiRise camera on the Mars Reconnaissance Orbiter. Large sand dunes border the wrinkle ridges, evidence of ongoing erosion and deposition by wind. On worlds such as Titan, Mars, and Earth, erosion has worked to reshape tectonic features. However, on dry worlds such as Mercury (inset), new images from the Messenger spacecraft reveal ridges and craters only affected by bombardment from space. Similarly, the lack of water on Venus results in relatively unchanging ridges and faults. Without information on the interior structure of Venus, we are still in the early stages of understanding the tectonics of this planetary neighbor.

Highs and Lows Plate motion on Earth brings continental crust into collision, crumpling and piling it into immense mountain ranges. Over time, erosion, mostly through the action of running water, reduces the jagged mountains to rounded hills. Such is the story of the Appalachians, whose low rounded summits (inset) seem to have little in common with the towering, rugged peaks of the Himalaya (large image). But these are just mountain ranges in different stages of development—the Appalachians are an old, eroding mountain range, with the Alps and Himalaya still growing upward. The force that formed the Appalachians was an ancient plate collision about 680 million years ago. At one time, the Appalachians were the central range on the supercontinent of Pangaea. Multiple plate collisions over eons continued to shape the mountain range, steadily reduced by erosion, and then pushed up again. The last mountain-building event took place more than 30 million years ago, lifting up ancient deformed rocks that have subsequently been deeply cut by streams and rivers.

Mountains in the Solar System

OUNTAIN RANGES are rarer in the rest of the solar system than on Earth, with Venus having the most impressive peaks among the other planets. Mars lacks any real tectonic-built mountains at all, as does Mercury—the only terrains that resemble mountains on those bodies are the rugged rims of impact craters. Two types of mountainous terrain have been seen on Venus: ranges similar to those on Earth and complex, deformed terrain called tessera.

Most of the mountain ranges on Venus ring a high plateau called Lakshmi Planum located in Venus's northern hemisphere. The plateau of Lakshmi rises about a mile above the rolling plains of Venus, and is ringed by mountain belts. The flat surface of the plateau is covered by volcanic flows. Unlike Earth's Tibetan plateau, Lakshmi was probably pushed upward by a hot plume of material from the interior of Venus, rather than resulting from plate collisions. With time, the plateau may have begun to spread outward under the force of gravity, causing some compression at its edges. Scientists are still puzzled by the forces that formed this region.

The highest mountain belt on Venus lies to the east of the plateau. Maxwell Montes rises almost 7 miles above the average surface elevation of Venus, about 4 miles above the surface of the plateau, and is about 530 miles long by 435 miles wide. The mountain range is made up of parallel ridges and valleys, cut by later extensional faults. The extreme height of Maxwell in comparison to the other compressional mountain ranges around Lakshmi suggests that its origin is more complex. Some scientists have even suggested that Lakshmi represents a remnant formed by an earlier episode of plate tectonics on Venus. The only way to really resolve the origins of Maxwell is to measure the composition of its rocks using robotic surface landers, and to investigate the structure of the lithosphere in this area by sending seismometers to the surface that would use energy from "venusquakes" to reveal subsurface details. Both missions pose technological challenges, given the extreme temperatures and pressures on Venus and the rugged landscapes of the Maxwell region.

SOLAR SYSTEM'S TALLEST MOUNTAINS

Tallest Feature	Planet	Height (miles)
Olympus Mons	Mars	15.5
Maxwell Montes	Venus	6.8
Mount Everest	Earth	5.4
Verona Rupes	Miranda (Uranus)	3
Huygens Mons	Moon (Earth)	2.9

To the east of Maxwell lies a region of terrain called Fortuna Tessera. When Russian scientists in 1983 first saw radar images from their Venera orbiters of the intersecting sets of fractures and ridges in this region, they called the feature tessera because the terrain reminded them of complex tile (tessera) floors. Along with Fortuna, there are six other large tessera regions, all of them plateaus ranging from about a half mile to a mile high. Scientists argue over whether tesserae formed over hot spots (regions of molten rock below and within the lithosphere), causing the surface to rise up and then collapse after the hot spot went away, or whether tesserae formed over areas of sinking mantle, which dragged the crust inward into a deformed pile. All of the regions of tessera on Venus appear to be relatively old; they are all surrounded by younger low-lying plains made up of multiple layers of volcanic flows. The tessera terrain may also preserve an earlier period in the geologic history of Venus, when the crust was moving more. It will take future spacecraft missions to Venus's surface to resolve this controversy.

Belts of ridges narrower than the mountain ranges or tessera regions are also found on Venus, rising to about a half mile above the surrounding volcanic plains. They form by much more limited horizontal motion of the surface, probably over areas where convective cells are descending into the interior of Venus, dragging the crust into ridges.

Venus's Tangled Tesserae Alpha Regio, a topographic highland on Venus approximately 790 miles across, is displayed in this three-dimensional perspective view (large image), produced by radar images from the Magellan spacecraft. The orange color mimics the filtering effects through the thick Venusian atmosphere. Alpha Regio is a mountainous region with multiple sets of intersecting trends of ridges, troughs, and flat-floored fault valleys that resemble mountainous regions like those in Yemen (inset) here on Earth. Directly south of Alpha is a large oval feature named Eve. The radar-bright spot located centrally within Eve marks the location of the prime meridian (zero longitude line) of Venus.

Earth's Motion

WRINKLES, RIDGES, AND FAULTS are scattered across the solar system: What are they telling us? By looking at all of these features, scientists are trying to unravel how a planet is layered, what these layers are made up of, and why and when they break. As with many things in science, trying to answer these tectonic questions boils down to math. Geophysicists measure the strength of rocks under different conditions in laboratories and combine this information with mathematical models. The models enable scientists to experiment on a planet-wide scale: What kind of faults would form if one pulled on a six-mile-thick layer of granite crust? How long would the faults be and how far apart would they occur? Scientists can compare these model-produced answers to what they see on the surface of Earth or Venus and start to decipher what those observed features tell us about the composition and structure of the upper layers of the planet. The advantage of having multiple planets is that each has a unique set of materials, deforming under slightly different conditions. For example, a basaltic rock deforming on Earth will not behave the same way as a basalt deforming on hot, dry Venus, where the lack of water makes rocks much stiffer and stronger. Getting a model to work under different planetary conditions—and produce sensible answers that match what we see—helps us improve the models' accuracy. Ultimately, this leads to better models to apply to Earth. Understanding in detail how planetary surfaces deform will, scientists hope, help us forecast surface movement and earthquakes. Several million earthquakes occur on Earth every year. Scientists can identify regions where earthquakes are more likely to occur, such as the faults ringing the Pacific Ocean, and they can study the history of a particular fault to forecast when another damaging quake might occur. However, the exact timing and magnitude cannot be predicted even for well-studied faults like the San Andreas, and often devastating earthquakes occur on unknown or not-well-studied faults, such as the 1994 Northridge California earthquake.

Mountain ranges and an abundance of rifts and faults are common features on planets and satellites in the solar system. But when it

Earthquakes are a hazard in highly populated regions such as Japan (left) and China (above). The 2008 7.9 magnitude earthquake in Sichuan, China, caused extensive damage and killed more than 68,000 people.

comes to studying the process that really controls how our own planet deforms—plate tectonics—we find no easy parallels: Earth appears to be the only world where plates grow, shift, and dive back into the mantle. This knowledge of tectonics is very model dependent, as many of these processes, such as mountain-building, are driven from deep within the Earth and take place on long timescales, where we can't watch the action in real time. However, those extraterrestrial mountain ranges, rifts, and faults are pieces of a crustal motion puzzle that helps to expand our understanding of tectonics. The planet with the most Earthlike tectonics is Venus, where we find many of the features, including mountain ranges, to aid comparisons and refine our models, though we still have a great deal to learn about Venus's tectonics. Earth's tectonics are unique, and unlocking those secrets awaits the marriage of our ongoing curiosity with our ability to build tougher, more capable robot geologists.

CHAPTER 4

FIRE FROM WITHIN:
VOLCANOES

Spectators witness a dramatic eruption of lava, ash, and gas from Sicily's Mount Etna.

Deadly Drama

VOLCANOES ARE ONE of the most common features on the surfaces of the planets and satellites of our solar system. Their lava flows reshape planetary landscapes, and their eruption plumes alter atmospheres. Volcanoes on Mars, Venus, Titan, Io, and Enceladus have dramatically shaped those planets and satellites. On Earth, volcanoes create dramatic scenery, from island paradises like the Hawaiian archipelago to towering, snow-capped peaks like Mount Fuji. Majestic as they are, however, volcanoes pose a hazard; they damage and destroy property with explosive clouds and lava flows and alter the atmosphere by injecting gas and ash. In order to understand the hazards, volcanologists compare Earth volcanoes to those on planets such as Mars and Venus. Those studies help us better determine the factors that control why, how, and most important, *when* a volcano will erupt. Our studies of Earth's volcanoes, in turn, help us understand the contemporary surfaces of other planets and how they have changed over time.

In often dramatic displays of erupting ash and lava, volcanoes reveal how—and where—a planet expels its internal heat. The internal heat, which comes from the decay of radioactive elements locked within rocks as well as heat left over from the formation of the planet, creates melted rock, or magma. This magma is less dense than the surrounding rock and rises to the surface. Gases such as water and carbon dioxide bubble within the magma and help to drive it upward, like soda from a shaken can, causing it to erupt on the surface.

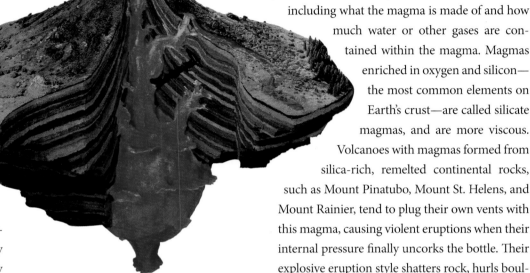

Layered lava flows, sometimes alternating with layers of volcanic ash, build a volcano's mountainous shape. A central vent carries magma from the underlying pool in the magma chamber to the top of the volcano.

Many factors affect the style of a volcanic eruption, including what the magma is made of and how much water or other gases are contained within the magma. Magmas enriched in oxygen and silicon—the most common elements on Earth's crust—are called silicate magmas, and are more viscous. Volcanoes with magmas formed from silica-rich, remelted continental rocks, such as Mount Pinatubo, Mount St. Helens, and Mount Rainier, tend to plug their own vents with this magma, causing violent eruptions when their internal pressure finally uncorks the bottle. Their explosive eruption style shatters rock, hurls boulders, and sends extensive clouds of ash into the atmosphere. This type of continental mountain, built up of layers of lava and explosive ash deposits, is called a composite volcano.

Less silica-rich lavas are much more fluid; gases escape easily and flows harden into a hard, black rock called basalt. On Earth, basaltic volcanoes on islands such as the Galápagos and Réunion erupt relatively gently, with long, fast-moving lava flows and fountains of fluid lava jetting from cracks or fissures. These volcanoes are called shield volcanoes because of their shallow, domed shape, which is built up of many eruptions of basaltic lava. The shield volcano is the most common type of volcano around the solar system.

Volcanologists study what goes on beneath and inside a volcano, as well as analyzing the flows, ash deposits, and gas plumes that erupt. Their insights have given the scientists who study volcanoes on other worlds a head start on understanding one of the solar system's most exciting phenomena.

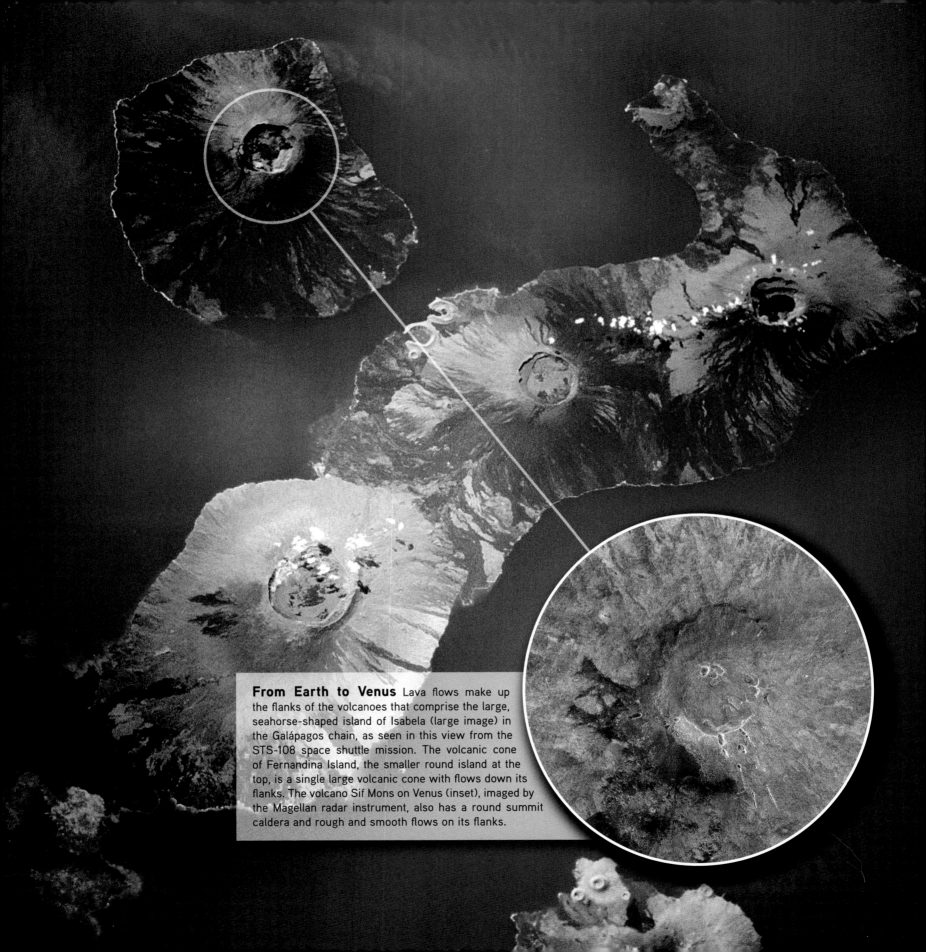

From Earth to Venus Lava flows make up the flanks of the volcanoes that comprise the large, seahorse-shaped island of Isabela (large image) in the Galápagos chain, as seen in this view from the STS-108 space shuttle mission. The volcanic cone of Fernandina Island, the smaller round island at the top, is a single large volcanic cone with flows down its flanks. The volcano Sif Mons on Venus (inset), imaged by the Magellan radar instrument, also has a round summit caldera and rough and smooth flows on its flanks.

Where Volcanoes Form

VOLCANOES HAVE BEEN FOUND ALL over the solar system, but very few planets still have active ones on their surface. The interior of a planet has to be generating enough heat to melt rock, and that magma has to be able to reach the surface through cracks in the outer solid crust of the planet. On most planets, that internal heat is produced by the decay of radioactive elements such as uranium and thorium. On some satellites, such as Jupiter's moon Io, internal heat is generated by gravitational stresses caused by the giant parent planet as its tides flex the moon. Some planets had a lot of internal heat early in their lives, but they did not have enough radioactivity to sustain volcanism over 4.5 billion years. On bodies such as Mercury and the moon, we can find evidence of past volcanism, but no current activity persists. The larger inner planets, Earth and Venus, have abundant supplies of radioactive elements and are thus more geologically active. Mars falls in the middle; most heat-driven geologic activity on Mars probably ceased millions of years ago.

Where you find a volcano on a planet tells you a great deal about how that world works internally. On Earth, volcanoes cluster at plate boundaries. At the boundary between spreading oceanic plates, volcanoes form in a long chain over hotter material upwelling from deep within Earth. This process forms the volcanoes of Iceland, which sit directly upon the Mid-Atlantic Ridge. Volcanoes also form over subduction zones, where the plate that descends into the interior of Earth melts, sending magma upward. The volcanoes of Japan and the Cascade Range of the western United States are subduction zone volcanoes.

Volcanoes can also form over places on planets called hot spots. Heat from the hot core of the planet can produce a plume of upward-moving melt, which penetrates the surface and produces volcanoes. On planets with moving plates, the plate slides across the hot spot, creating a chain of volcanoes such as the Hawaiian Islands. Hot spots can also form beneath continents; Yellowstone is an example of a continental hot spot volcano. It produced the biggest North American eruption in the last million years, burying much of the northern Great Plains under yards of volcanic ash.

Earth is the only body in the solar system where plate tectonics is known to operate: No other planet has the distinctive chain of volcanoes that form at plate boundaries. On other planets or moons, volcanoes are scattered more randomly across the surface. Scientists cannot rule out that plate tectonics did occur at some point early in the history of Mars and Venus, but we see no current evidence of it. Volcanoes on Mars and Venus do form at hot spots, such as the volcanoes of the Tharsis rise on Mars. Hot spots are also thought to form volcanoes on Jupiter's moon Io.

All formed by volcanism, the seven islands of Hawaii are visible in this Earth Observing System satellite image. From left to right: Niihau, Kauai, Oahu, Molokai, Lanai, Kahoolawe, Maui, and Hawaii. On land, only Hawaii island's Mauna Loa and Kilauea are currently active.

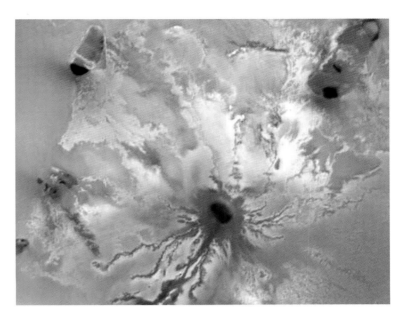

Ra Patera is a volcano on Io, one of the moons of Jupiter. Io is the most active volcanic body in the solar system, with more than 70 erupting volcanoes. On Earth, approximately 20 volcanoes are erupting at any given time, and around 60 erupt every year.

Volcanism on Earth occurs in three main settings: at subduction zones (left), at spreading centers (center), and over hot plumes of material that rise from deep in the interior (far right).

Namesakes Volcanic eruptions can last anywhere from days to weeks to years, with a typical eruption lasting about seven weeks. The types of eruptions are named after volcanoes that erupt in that fashion—for example, a strombolian eruption is intermittent fountains of lava, named after the Italian volcano Stromboli. Other types, from least explosive to most explosive, are hawaiian, vulcanian, vesuvian, pelean, and plinian.

Blast From the Past

ACROSS THE COURSE of human history, large volcanic eruptions have wreaked havoc, killing thousands and even disrupting the course of civilizations (as on the island of Crete, more than 3,000 years ago). Scientists have tried to rate volcanic eruptions, based on factors such as the amount of material erupted and the height reached by the eruption plume. One such scheme rates eruptions on a scale of zero to eight, with eight being the most destructive. Fortunately, over the last 10,000 years, there have been no eights and only one seven: the 1815 eruption of Tambora. The eruption of Tambora, on the island of Sumbawa in Indonesia, was so destructive that it killed 92,000 people and caused the global climate to cool, producing the Year without a Summer in 1816. This kind of eruption occurs about every thousand years. Krakatau in 1883 rates a six, and the eruption of Mount Vesuvius that destroyed Pompeii rates a five. The latter's eruption in A.D. 79 is rated just slightly higher than the 1980 explosion of Mount St. Helens. However, Vesuvius was far more destructive: The local area was populous, and the eruption's fast-moving cloud of hot gas and ash buried the

DANGEROUS VOLCANOES

These are some of the Decade Volcanoes—volcanoes designated for study and monitoring by the International Association of Volcanology and Chemistry of Earth's Interior (IAVCEI), as they pose multiple hazards to populated regions and have been recently active.

Volcano	Location
Avachinsky-Koryaksky	Kamchatka, Russia
Colima	Jalisco, Mexico
Mount Etna	Sicily, Italy
Galeras	Colombia
Mauna Loa	Hawaii, U.S.A.
Mount Merapi	Central Java, Indonesia
Mount Nyiragongo	Democratic Republic of the Congo
Mount Rainier	Washington State, U.S.A.

A.D. 79: The devastating eruptions from Vesuvius buried Pompeii, taking lives and destroying the town. Quiet since 1944, Vesuvius is not dead. The volcano (above) sits on the edge of the city of Naples, Italy, and could produce dangerous flows of hot ash that destroy everything in their path. Ongoing, careful monitoring of Vesuvius is essential.

surrounding towns. The region around Vesuvius is now even more highly populated; the mountain is surrounded now by more than three million people in Naples, Italy, and remains active (last erupting in 1944). Enormous scale eight eruptions have occurred in the distant past, such as the giant explosions that formed the Long Valley caldera in California and the Yellowstone caldera. The Long Valley eruption occurred about 730,000 years ago and deposited volcanic ash as far away as Kansas. The even larger eruption of Toba in Sumatra about 75,000 years ago sent about 20 times more ash and dust into the atmosphere than the eruption of Tambora, and likely produced even more dramatic climatic effects. Luckily, extremely large eruptions are very infrequent, probably occurring about once every five million years.

Do large volcanic eruptions contribute to global climate change today? Volcanoes release about 117 million tons of carbon dioxide, the main greenhouse gas, into the atmosphere every year. That may seem like a lot, but it pales in comparison to almost 5.5 billion annual tons of human-produced carbon dioxide emissions, making volcanoes a small player in global warming.

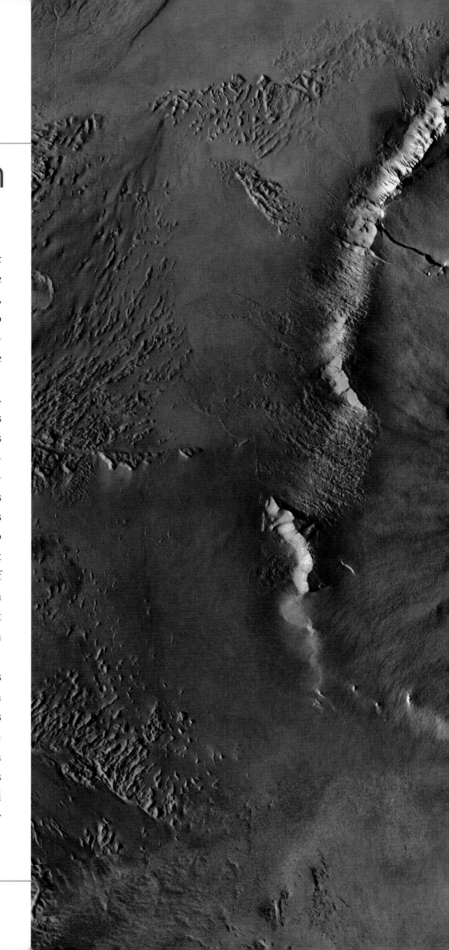

Large Volcanoes of the Solar System

LARGE VOLCANOES are some of the most dramatic geologic features in the solar system. They are built flow by flow, with more frequent, smaller eruptions combining with the output of rarer, large eruptions. Eruptions originate both at the summit of the volcano and from vents along its flanks. Often, dramatic calderas (large depressions) form at the summit of volcanoes, caused by the collapse of the mountaintop as magma drains out of the chamber within the volcano.

The largest volcano in the solar system is Olympus Mons on Mars. Towering 15 miles above the Martian surface, it is more than 370 miles across. It would cover nearly the entire country of Poland. Olympus Mons sits over what was probably a Martian hot spot, but without plate tectonics on the red planet, the volcano just kept growing and growing without producing the chain of volcanoes seen at hot spots on Earth. It was also very long-lived, active for as long as a billion years. The volcano has grown so large that it has collapsed along its margin, producing a cliff up to 3.6 miles high. This is not unusual—volcanoes often grow so tall that just the force of gravity is enough to make them become unstable. Part of the flank of Mount Etna in Sicily has collapsed, leaving a large depression called the Valle del Bove. Olympus Mons is unlikely to be active; most volcanism on Mars probably ceased millions of years ago, as can be seen by the cratered (and thus relatively ancient) lava flows on its flanks.

Large volcanoes on Mars occur in two main groups: the Tharsis group that lies to the southwest of Olympus Mons and the Elysium group. The three volcanoes in Tharsis—Arsia, Pavonis, and Ascreus Montes—sit at nearly the same height as Olympus Mons but are actually less massive, as they are perched on a very thick pile of broad lava flows. Age dating of the flows of these volcanoes using impact craters seems to indicate that volcanic activity may have shifted north toward Olympus Mons; even on a planet without plate tectonics, hot spots may shift in position over time.

Forged by Fire The rocks produced by volcanoes like Mars's Olympus Mons (large image) or during the eruption of Emi Koussi in Chad (inset) are igneous, formed from the cooling of lava. Igneous rocks also form under Earth's surface when magma cools before reaching the surface. There are two other types of rocks in the solar system—sedimentary rocks, which are made up of sediments such as mud and gravel that have been hardened into rock; and metamorphic rocks, which are igneous or sedimentary rocks that have undergone high temperatures and pressures.

Extreme Science Studying features like Maat Mons on Venus (large image) is a lot more difficult than exploring Mauna Loa in Hawaii here on Earth (inset). With 900°F and 90 times the atmospheric pressure of Earth, landers on Venus can survive only a few hours. Landers sent by the Soviet Union survived long enough to take pictures of the surface and measure rock compositions. Spacecraft orbiting Venus have studied the atmosphere and surface, using radar to image the surface. Radar waves penetrate through the perpetual cloud cover on Venus, returning images that look like photographs. In radar images, the variations in brightness give information about how rocky and rugged the surface is, and how much soil is on the surface.

Volcanoes that dwarf those on Earth are also found on Venus, which has more volcanoes than any other body in the solar system. The tallest volcano on Venus is Maat Mons, which rises some 4.8 miles high and reaches about 480 miles across. Maat may still be active—we just don't know; the radar instrument on the Magellan spacecraft that studied Venus in 1990-94 had no ability to detect an erupting volcano. However, an earlier mission to Venus had measured declining levels of sulfur dioxide in the atmosphere; that gas may have been produced by erupting volcanoes on Venus. So similar in size to Earth, Venus is likely to still be geologically active. Future missions to Venus will search for more concrete evidence of an erupting volcano.

Large volcanoes on Venus are scattered across the surface, with no clear pattern that would indicate the edges of tectonic plates. Many of the volcanoes occur in groups on top of topographic bulges, such as Sif and Gula Montes. At least ten such volcanic rises have been mapped on Venus, suggesting that hot spots may be as common on Venus as on Earth.

Even though the volcanoes of Venus and Mars are larger in scale than terrestrial examples, the processes that formed them seem to be very similar. The large shield volcanoes seem to build up in similar ways, with layer upon layer of lava flows. The shapes and textures of individual lava flows are very similar to those on Earth, and studies of terrestrial flows on Hawaii and at Etna have helped geologists understand the principles that control the flow of liquid rock across the solar system. But Venusian volcanoes are still surprising. In spite of the differences between dry, scorching Venus and temperate Earth, scientists nevertheless expected lava

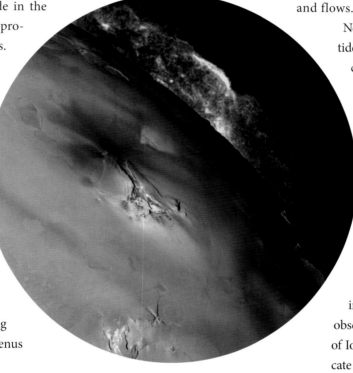

The volcano Pele erupts on Jupiter's moon Io, imaged by NASA's Galileo spacecraft. The plume from the volcano rises 180 miles above the surface, with material falling over an area the size of Alaska. Volcanoes on Io are surrounded by colorful lava flows rich in sulfur.

flows to be about the same length as those on Earth. They assumed that the lack of water to drive volcanic eruptions would offset the slower cooling of flows due to the high Venusian surface temperature. However, the Magellan spacecraft discovered that flows on Venus are very long, with some flows extending more than a hundred miles away from volcanoes. Volcanologists are using this new Venus data to help improve models of how lava forms and flows.

Next to Earth, Jupiter's moon Io, rent by tides from its massive captor, is the most volcanically active body in the solar system. The largest volcanic features on Io are paterae, Latin for "calderas," or craters, some occurring on large, low volcanic shields. The paterae are more than 180 miles across, with individual flows over 90 miles long. They are thought to be about 1.2 miles deep. Paterae like Pele are surrounded by multicolored sulfur and silicate flows. The sulfur in Io's lava flows produces its characteristic reddish yellow coloring, but the high eruption temperatures observed by the Galileo spacecraft and images of Io taken by telescopes here on Earth indicate that many of the erupted flows are silicates, some with very unusual compositions. The hottest lavas on Io seem to be similar to volcanic rocks that have not erupted on Earth in over a billion years. The Galileo spacecraft detected lava lakes inside these giant calderas, similar to the pools of molten rock sometimes found inside terrestrial volcanic craters.

The volcanoes of Io are so active that it is the only body in the solar system to have no identifiable impact craters on its surface. About 350 volcanoes have been identified on the surface; more than 70 are thought to be active. Some mountains and ridges are seen on the surface, but whether they are the result of volcanism or tectonics is not clear.

Catching a Volcano in the Act

AT ANY GIVEN TIME, as many as 10 to 20 of Earth's approximately 600 active volcanoes are erupting. Kilauea volcano in Hawaii has been erupting at a nonviolent but steady pace since 1983, and descriptions of eruptions at Mount Etna date back to the ancient Greeks, who identified Etna as the forge of Hephaestus (the Roman god Vulcan), the god of fire, and home to the giant race, the Cyclopes. On Earth, volcanic eruptions are monitored by an array of devices, from tilt meters and seismometers that detect magma moving underground to sensitive instruments that measure changes in venting volcanic gases. Satellites also detect heat emissions from volcanoes in remote areas of Earth and photograph their ash plumes. On other bodies, it is much harder to catch a volcano in the act of erupting, especially on a planet like Venus that is covered in clouds.

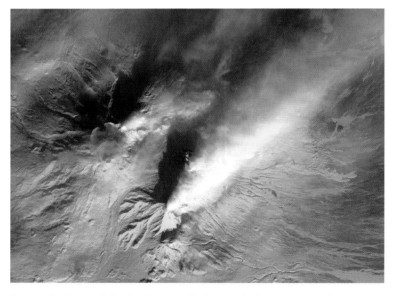

Caught in the act! Jupiter's moon Io is caught in an extraordinary mid-eruption image (right) by NASA's Galileo spacecraft. NASA's ASTER satellite caught another eruption of Kliuchevskoi volcano (above) in 2005. This eruption sent a lava flow down the northern flank of the volcano that melted some of the glaciers that surround it.

From orbit around Earth, the most easily identifiable sign of a volcanic eruption is the plume of ash arising from the summit of an active volcano. These plumes, which consist of steam and volcanic ash, rise vertically up into the atmosphere and then move horizontally, pushed by prevailing winds or the jet stream. Astronauts on the International Space Station caught the 2002 eruption of Mount Etna, which produced an ash column that rose more than three miles. These plumes of ash can reach up to 12 miles into the atmosphere.

Major volcanic eruptions produce clouds of materials that can spread all over the planet in a matter of days. The ash produced by Pinatubo in the Philippines in 1991 circled the globe in the stratosphere, producing spectacular sunrises and sunsets. Shuttle astronauts reported that the atmosphere stayed noticeably dusty for more than a year.

Lava flows—on Earth and beyond—are harder to detect from space. Scientists can use sensitive instruments that detect changes in surface temperature to detect eruptions, or monitor how a volcano changes in appearance over time. The shuttle's Space Radar Laboratory tracked the progress of Kilauea's lava flows in Hawaii by studying the flows on missions six months apart in 1994. Satellites orbiting Earth also monitor volcanoes, using instruments that measure the temperatures, compositions, and motions of lava flows. Using a technique called radar interferometry, motions of the surface as little as inches in extent can be measured.

In 1979 the Voyager 2 spacecraft flew by Io, photographing this brightly colored world for the first time. Scientists concluded the fresh-appearing surface, with no visible impact craters, must be very young. An engineer studying images of Io's limb, or outer edge, for navigation purposes noticed a crescent-shaped cloud of sulfur compounds extending above the moon, which has only a very tenuous atmosphere. The image (at right) had captured an eruption in progress. Scientists quickly found other examples of plumes rising off the surface and realized that the Io volcanoes were extremely active. The Galileo spacecraft, which imaged Io from 1995 to 2001, repeatedly monitored the eruptions of Io's volcanoes and even captured lava fountains erupting from surface vents.

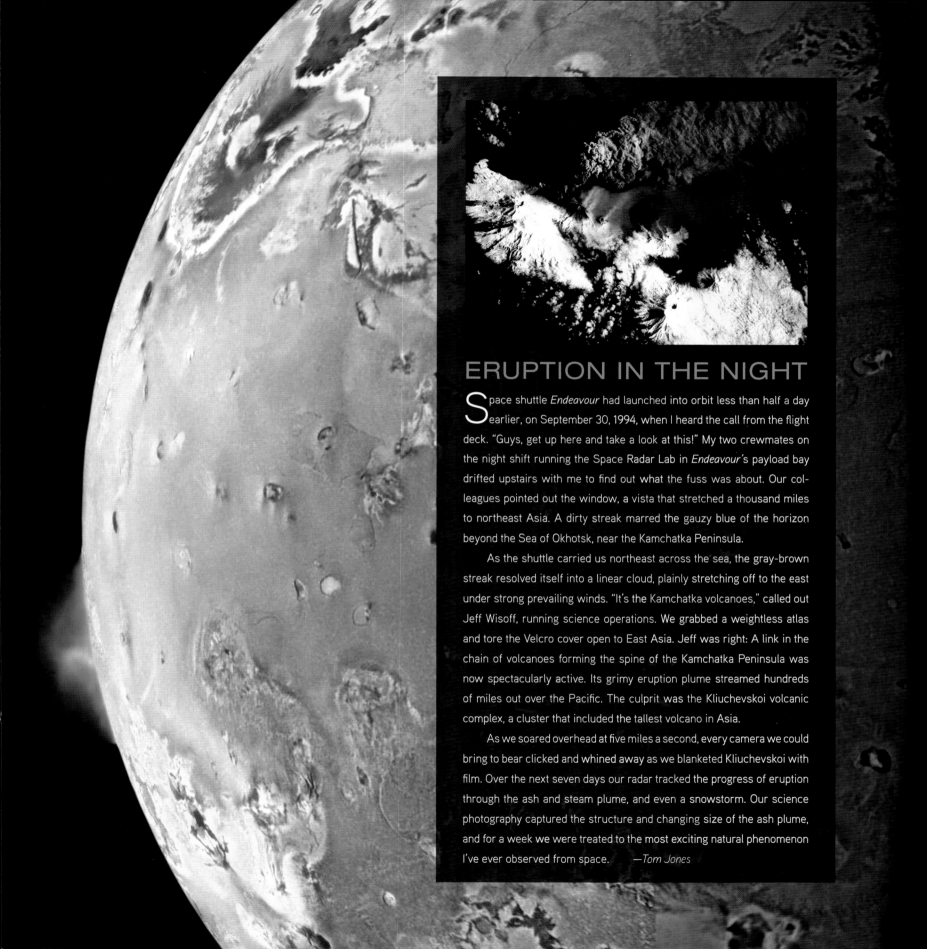

ERUPTION IN THE NIGHT

Space shuttle *Endeavour* had launched into orbit less than half a day earlier, on September 30, 1994, when I heard the call from the flight deck. "Guys, get up here and take a look at this!" My two crewmates on the night shift running the Space Radar Lab in *Endeavour*'s payload bay drifted upstairs with me to find out what the fuss was about. Our colleagues pointed out the window, a vista that stretched a thousand miles to northeast Asia. A dirty streak marred the gauzy blue of the horizon beyond the Sea of Okhotsk, near the Kamchatka Peninsula.

As the shuttle carried us northeast across the sea, the gray-brown streak resolved itself into a linear cloud, plainly stretching off to the east under strong prevailing winds. "It's the Kamchatka volcanoes," called out Jeff Wisoff, running science operations. We grabbed a weightless atlas and tore the Velcro cover open to East Asia. Jeff was right: A link in the chain of volcanoes forming the spine of the Kamchatka Peninsula was now spectacularly active. Its grimy eruption plume streamed hundreds of miles out over the Pacific. The culprit was the Kliuchevskoi volcanic complex, a cluster that included the tallest volcano in Asia.

As we soared overhead at five miles a second, every camera we could bring to bear clicked and whined away as we blanketed Kliuchevskoi with film. Over the next seven days our radar tracked the progress of eruption through the ash and steam plume, and even a snowstorm. Our science photography captured the structure and changing size of the ash plume, and for a week we were treated to the most exciting natural phenomenon I've ever observed from space.   —Tom Jones

Volcanoes of All Sorts

THE LARGE volcanoes described previously are all shield and composite volcanoes, most with the classic cone shape of Japan's Mount Fuji. However, volcanoes come in all shapes and sizes. Smaller volcanoes range in shape from low-lying shields to cones to flat-topped domes, with the differences in morphology reflecting differences in lava composition and eruption style. By looking at the shapes of volcanoes and the shapes and textures of lava flows, volcanologists can begin to understand what type of rock makes up the volcano and how it formed.

Volcanic features commonly occur in groups, fed by a long-lived magma chamber beneath the surface. On Earth, a good example of a volcanic cluster is the Cima volcanic field in California. The Cima field covers about 90 square miles and has more than 65 lava flows and 52 volcanic vents. On the moon, groups of small volcanoes have also been found in the broad lava plains that fill large impact basins (the lunar maria). The Marius Hills, some 90 miles across, is the largest cluster of volcanic domes and cones on the moon. It was the alternate landing site for Apollo 15, but was not selected for a manned expedition. Most of the groups of volcanic domes on the moon are ancient; the Marius domes and cones are older than the surrounding lavas, which are several billion years old.

Clusters of volcanoes are also common on Venus and Mars. On Venus, some areas, called shield fields, contain hundreds of small volcanoes and often cover areas so large—up to 300 miles—that the fields must be underlain by a broad plume of hot material. Fields of cones on Mars may be telling us something about water under the surface. One field of cones—described as rootless cones—contains volcanic features that form

Resembling a river valley, Hadley Rille, on Earth's moon, extends for more than 75 miles. Found on Venus, Mars, and the Moon, rilles are cut into the surface by flowing lava.

when an underlying layer of water or ice covered by a lava flow produces explosions. Such cones also form on flows in glacier-covered Iceland.

Lavas that are very rich in silica form very different types of structures than the basaltic shield volcanoes. As we have seen, silica makes lava very viscous, or sticky, producing explosive eruptions and flows of large blocks, like the 1980 eruption of Mount St. Helens. Finding such flows is of particular interest to volcanologists because they provide clues to the history of the crustal rocks of the planet—these silica-rich flows form when the crust of a planet gets recycled and melts again. On Mars and Venus, where plate tectonics does not occur, and there are few chances for crustal rocks to be subducted and melted at depth, silicic volcanoes are rarer. Mars has a few large volcanoes that may have produced explosive eruptions, including Tyrrhena Patera. This volcano is about 1.6 miles high and 180 miles across. The erosion seen around the volcano suggests that Tyrrhena was surrounded by explosive ash deposits, similar to those found at the Long Valley caldera in California.

Sinuous rilles are probably the strangest volcanic features found on other planets. First seen on the moon, sinuous rilles look like river valleys that often begin at a pit. Hadley Rille on the moon, which is more than 75 miles long, was visited by the Apollo 15 astronauts. It formed about three billion years ago. Rilles may form when a lava tube collapses, or when a very hot river of lava melts the underlying rock, cutting a channel. However, the moon's rilles are much larger than lava channels on Earth, which tend to be less than six miles long. A sinuous rille found on Venus is nicknamed "Gumby" because of its resemblance to the cartoon character.

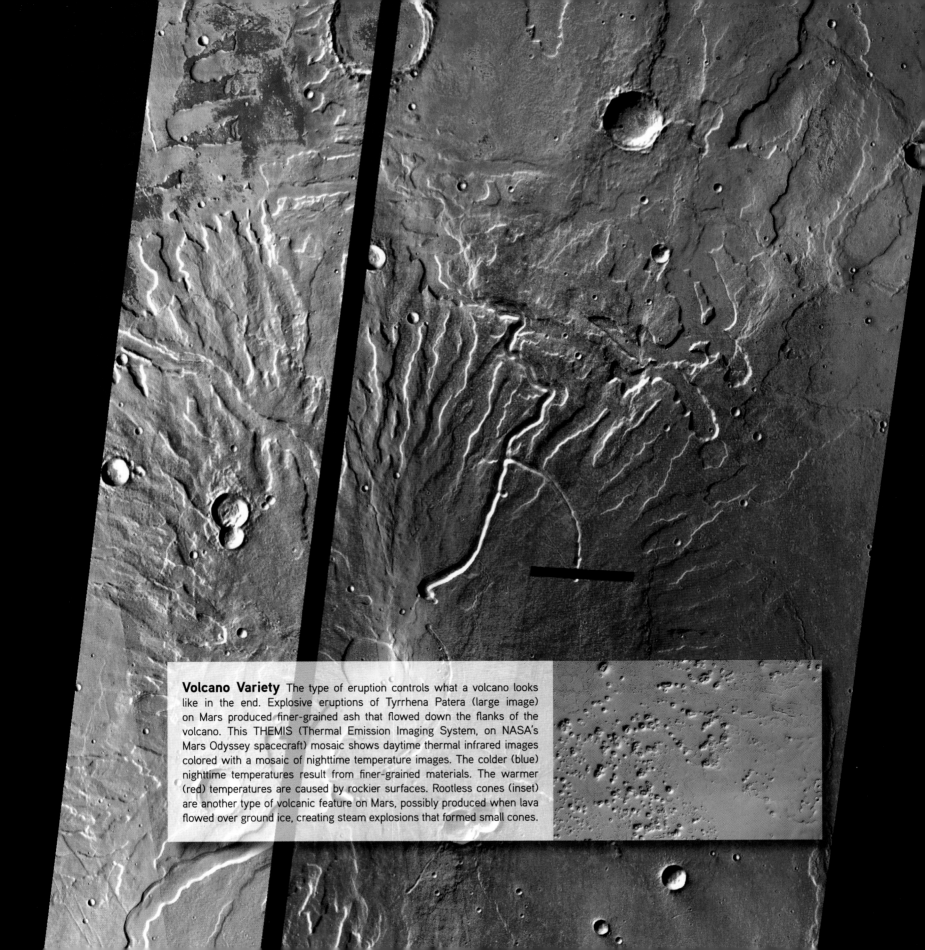

Volcano Variety The type of eruption controls what a volcano looks like in the end. Explosive eruptions of Tyrrhena Patera (large image) on Mars produced finer-grained ash that flowed down the flanks of the volcano. This THEMIS (Thermal Emission Imaging System, on NASA's Mars Odyssey spacecraft) mosaic shows daytime thermal infrared images colored with a mosaic of nighttime temperature images. The colder (blue) nighttime temperatures result from finer-grained materials. The warmer (red) temperatures are caused by rockier surfaces. Rootless cones (inset) are another type of volcanic feature on Mars, possibly produced when lava flowed over ground ice, creating steam explosions that formed small cones.

VOLCANOES

At Mount Etna in Sicily and in Hawaii, I have been studying how lava flows form and grow. We look at the final surface features of lava flows, and relate that to what we know about how a particular lava flow formed. Working at active volcanoes is always exciting, from watching glowing lava flows creeping across the ground at Kilauea in Hawaii to having ash rain down on your head as Mount Etna clears its throat once again.

My favorite volcano fieldwork has been done at silicic domes (an extruded plug of pasty, viscous lava), especially at the Inyo domes near Mammoth, California. When we saw the images of the steep-sided domes on Venus, everyone noted the similarities to steep-sided silicic domes on Earth. The radar images from Venus told us the surfaces of those domes were relatively smooth. In order to figure out how similar the Earth domes were to the Venus domes, we decided to measure the roughness of the surface of some silicic domes in California. The Inyo domes near Mammoth are a series of silicic domes that were erupted about a thousand years ago. We quickly learned that these surfaces are extremely rough; in fact, our measurements demonstrated that they are among the roughest surfaces on Earth. On a practical level, that meant lots of scraped legs from climbing over huge glassy lava boulders and a completely worn-out pair of hiking boots in a single field season! And then there was the too-close encounter with the bear....

Our work in the end showed that despite the similarities in shape, the Venus domes are unlikely to be silicic in composition: Their smooth surfaces indicate that they erupted in a manner very different from the eruptions of the terrestrial domes.   —*Ellen Stofan*

Shield Volcanoes Steep-sided, flat-topped domes on Venus (inset), may have formed by silica-rich eruptions, but they do not have the rough surfaces that silicic lavas produce at volcanoes like the Inyo domes in California (large image). On Earth, silica-rich domes like these often form inside the caldera of a large volcano after a major explosive eruption. Silica-rich domes are growing inside the calderas at Mount St. Helens in Washington State and in the Soufrière volcano on the island of Montserrat in the Caribbean. Intermittently between 1995 and 2007, Soufrière dome has been growing and partially collapsing to produce pyroclastic flows of hot ash and gas. These flows, also called *nuée ardentes* (glowing clouds), can be deadly.

Ice Volcanoes *of the Outer Solar System*

THE NATURAL SATELLITES of the outer planets are composed primarily of water ice, which was the most abundant material in that part of the solar system when they formed 4.5 billion years ago. Most of the satellites also incorporate some amount of rock, and thus have the ability to produce internal heat, though much less than the rocky bodies of the inner solar system. That internal heat, combined with that generated by the tidal pull of their parent planets, enabled some of the icy satellites of the outer solar system to become volcanically active. However, when these volcanoes erupt, they spout water mixed with other components, such as ammonia. This ice-rich eruption process, called cryovolcanism, is truly exotic, but it's volcanism all the same. The mechanisms of eruption of ice volcanoes are the same as their more familiar rock cousins. The behavior of water at the very cold temperatures on these outer solar system satellites is analogous to rock; melted ice erupts and flows just like the much hotter lavas on Earth.

At Jupiter, the moons Europa and Ganymede both show evidence of possible ice-rich volcanism. Bright, linear, grooved terrain cuts across the surface of Ganymede, flows made of almost pure water ice. This composition suggests that the bright terrain may have been formed by cryovolcanism, but most of it is heavily fractured and lacks any features typical of volcanism. Only a few depressions have been seen there that may be volcanic calderas. On Europa, which is thought to have a liquid layer beneath its outer ice shell, so-called band and ridge terrain may also provide evidence of cryovolcanism. Other small circular features and flowlike smooth deposits have also been seen that are more typical of volcanic features on other planets.

Twice as far from the sun, several icy moons of Saturn have extensive smooth areas that may have been resurfaced by cryovolcanic flows. Dione, Enceladus, Iapetus, and Rhea all show smooth-appearing areas in Voyager images that are now being investigated by the Cassini spacecraft. High-resolution images of Enceladus do show less cratered areas, and evidence of possible eruptions from long fractures.

Saturn's most intriguing moon is Titan, the second largest moon in the solar system with a radius of 1,600 miles; Jupiter's satellite Ganymede is slightly larger. Titan is the only moon in the solar system to have clouds and a thick, planetlike nitrogen atmosphere. The atmosphere also contains hydrocarbons like ethane and methane, which exist as liquids on the cold surface of Titan (-289°F). The Cassini spacecraft has a radar instrument that can see through the clouds and haze of Titan's atmosphere. The first images of the surface of Titan taken by the Cassini radar show a number of apparent cryovolcanic features, much to the surprise of the Cassini scientists who had expected a frozen, ancient surface covered with impact craters. Radar-bright lava flow-like features extend across the surface, and a bright-edged circular feature resembles the flat-topped volcanic domes of Venus. Titan is being revealed as a very geologically active world, as well as a possible analog to the primordial Earth.

Triton is the largest satellite of Neptune and has the coldest surface in the solar system (–391°F). The dark streaks on the south polar cap are deposits from geyserlike eruptions, and the area near the equator is thought to have been smoothed by cryovolcanic eruptions.

Saturn's Active Moons From the eruption plumes of Enceladus (large image) to the ice volcanoes of Titan (inset), Saturn's moons are amazingly active. NASA's Pioneer 11, Voyager 1, and Voyager 2 all flew by the Saturnian system, but Cassini-Huygens is the first spacecraft to orbit Saturn and study its atmosphere, its complex ring system, and its many moons. Seventeen nations aided in building the instruments of the orbiter and the experiments on the Huygens probe.

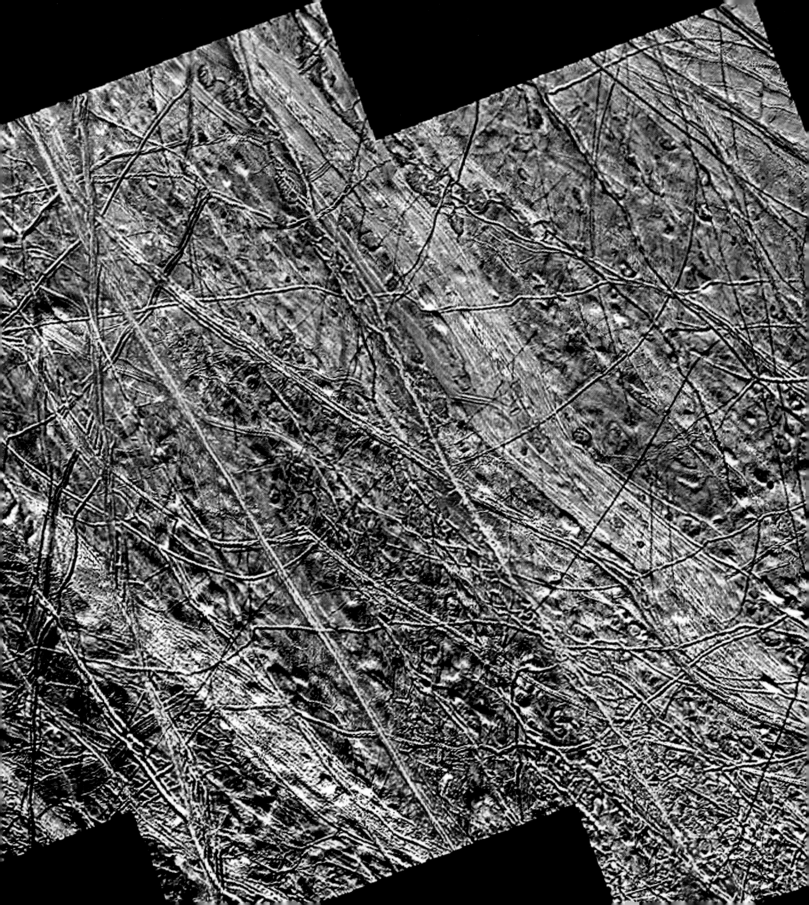

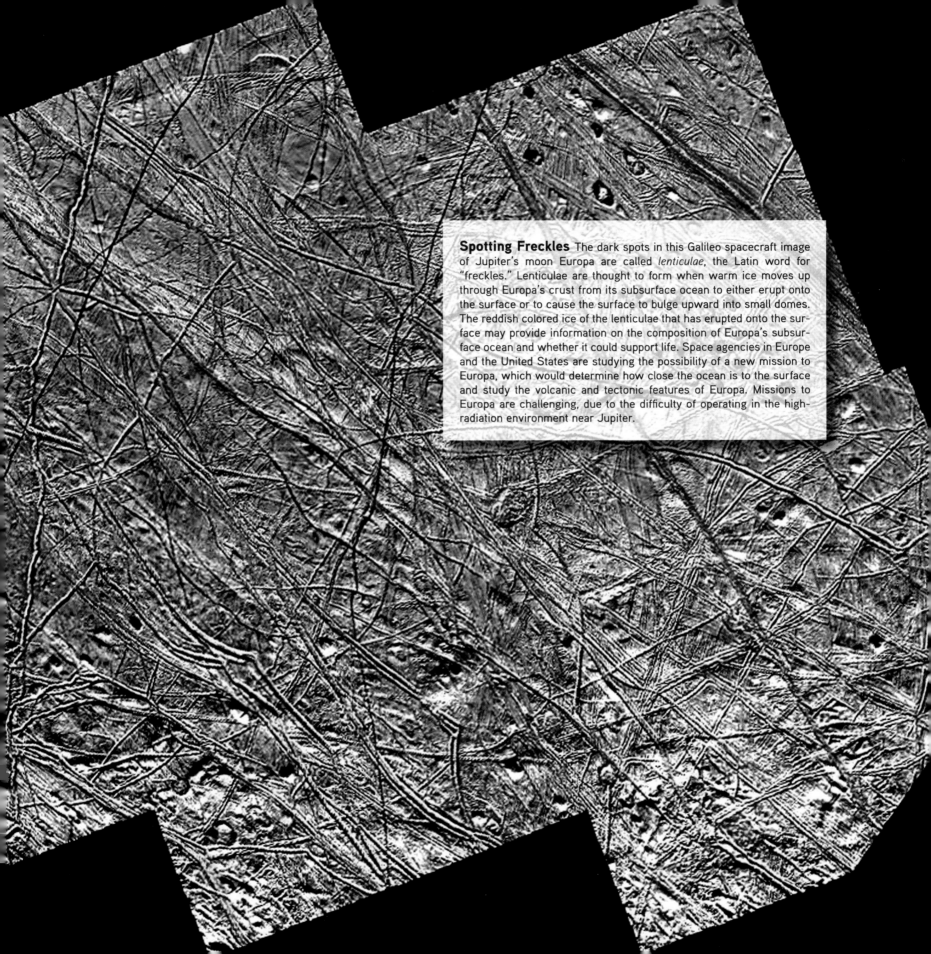

Spotting Freckles The dark spots in this Galileo spacecraft image of Jupiter's moon Europa are called *lenticulae*, the Latin word for "freckles." Lenticulae are thought to form when warm ice moves up through Europa's crust from its subsurface ocean to either erupt onto the surface or to cause the surface to bulge upward into small domes. The reddish colored ice of the lenticulae that has erupted onto the surface may provide information on the composition of Europa's subsurface ocean and whether it could support life. Space agencies in Europe and the United States are studying the possibility of a new mission to Europa, which would determine how close the ocean is to the surface and study the volcanic and tectonic features of Europa. Missions to Europa are challenging, due to the difficulty of operating in the high-radiation environment near Jupiter.

In the Shadow

THE DIVERSE ARRAY of volcanic features, from towering shield volcanoes to stumpy, rocky domes, tells us that volcanism is a very common process on many of the planets and satellites across the solar system. Volcanism provides a window into the interior of these bodies. The amount of volcanic activity reveals how internally active a planet is. The compositions of the lava erupted onto the surface, and the gases the lava contains, tell scientists about the chemistry of the interior of a planet. As long as a planet or moon has some amount of internal heat, internal melting can produce surface volcanism. That heat can come either from the decay of radioactive elements inside the planet or from a parent planet like Jupiter or Saturn tugging gravitationally on a moon and heating up its interior. The style of volcanism depends on the composition of the body—ice or rock—and the complexity of its geologic processes. Thus, the larger rocky planets such as Earth and Venus have an amazing assortment of volcanic landforms.

Somewhere on Earth, a volcano is erupting right now, causing people to evacuate—and possibly lose—their homes and businesses. In other places, like Naples, Italy, and Seattle, Washington, residents eye their local volcanoes (Vesuvius and Mount Rainier, respectively) with trepidation, knowing they have caused destruction in the past and wondering when the next eruption will occur. Along with Venus and Jupiter's moon Io, Earth is a very volcanic place, with plate tectonics and hot spots causing volcanoes to form on all of the continents and in all of the ocean basins. Volcanologists study the variety of volcanoes across the solar system to understand how high eruption plumes of ash will grow and spread, how these eruption plumes can affect the climate, how far lava flows will go, and how fast they will move. We can never hope to stop a volcanic eruption, but we can learn how to better prepare to prevent the loss of life and minimize damage to property. All of these volcanic processes can have a major impact on human life, and it will take the full array of our solar system's volcanoes to help us understand and eventually predict the behavior of these awesome but destructive features.

A Looming Threat Kirkjufell or Eldfell volcano (large image) erupts above the town of Vestmannaeyjar, Heimaey Island, Iceland. The 1973 eruption of this volcano threatened the port town, causing much of the population to evacuate. For five months, seawater was pumped onto the erupting lava flows to stop them from filling up the harbor. The eruption gradually died down and the harbor was saved. For some, like these children playing near the steaming Popocatepetl volcano in Mexico (inset), growing up in the shadow of a dangerous volcano is simply a way of life.

FLUIDS IN MOTION

The mighty Brahmaputra River drains the Himalaya and waters the fertile lowlands of India and Bangladesh, where it merges with the Ganges to form the world's largest delta.

Water in the Solar System

SEE ALSO: *Europa's Watery Secrets* 180

EARTH IS A WATER PLANET. Seventy percent of its surface is covered by oceans. All that H_2O gives our world the deep blue tint seen by robotic spacecraft and marveled at by lunar explorers 40 years ago. Earth's dry land bears the inescapable evidence of the force of water on its inexorable downhill passage to the seas. That evidence is impossible to ignore from space, and on the ground the handiwork of water is all around us. The Grand Canyon, today one of America's most spectacular national parks, provides dramatic evidence of water's power. The Grand Canyon wasn't formed by an earthquake or the opening of tectonic faults. Carving its way to the Sea of Cortés, the Colorado River sliced through a mile of rock spanning more than two billion years of Earth's geologic history, in the process conveying billions of tons of sediment into the oceans. If you've ever camped in the Grand Canyon, lying down each night next to the Colorado's rushing waters, drifting to sleep on Rocky Mountain sands thrown up by the river, its steady murmur has no doubt assured you that it is still at work, grinding ever deeper into its confining bed.

Earth, awash in its oceans 93 million miles from the sun, is the showcase for the power of water in shaping a planetary landscape. Water is a plentiful molecule throughout the solar system, but only on Earth does it exist stably in all three forms—gas, solid, and liquid. Venus may once have possessed an ocean, but its runaway greenhouse atmosphere has long since sizzled away any water at its surface. Closer to the sun, at Mercury, broiling temperatures instantly flash water into steam, which either escapes into space or breaks down under solar radiation into free oxygen and hydrogen (most of the latter inevitably escapes into space). At Mars and beyond, temperatures are too cold for liquid water to exist at the surface, or pressures are so low (or nonexistent) that any liquid water instantly vaporizes. Water is plentiful in the outer solar system, in comets or on dozens of satellites, but only as ice, which remains stable at low temperatures for millions of years.

Earth's atmosphere supports a hydrologic cycle where liquid evaporates into vapor but can cool enough in the atmosphere's upper layers to condense again into liquid and fall as rain or snow, eventually returning to rivers, lakes, and aquifers. Water not lost to evaporation flows down to the planet's seas and oceans, where an average water molecule resides for about 3,200 years. Evaporation from the oceans begins the process again. That perpetual cycle, driven by the sun's warmth, is what animates the erosive hand of water across the face of the continents.

Life flourished in Earth's oceans, as exemplified by this ammonite fossil. Related to the modern squid, ammonites thrived in ancient seas.

TITAN LAKES

Water is the most powerful sculptor of Earth's varied landscapes. Cycling from condensation to precipitation, it endlessly shapes and reshapes the land, as demonstrated by the action of Utah's Dirty Devil River, a tributary of the Colorado River (large image). Saturn's moon Titan is also shaped by fluids. Dark areas near Titan's north pole (seen in a Cassini radar image) are hydrocarbon lakes—some larger than Lake Superior. An active "methane-ologic" cycle operates on Titan—not with water but with methane (our "natural gas"), which comprises 6 percent of Titan's atmosphere. Methane evaporates from the surface and some falls again as rain. Other organic compounds, like ethane, propane, and sootlike solid particles, form as sunlight breaks down methane vapor. Scientists suspect this liquid hydrocarbon "precipitation" erodes the icy surface, percolates underground, or collects in lakes; warmer temperatures evaporate the methane to start the cycle anew.

A Powerful Force

A T THE HIGHEST ELEVATIONS on Earth, in the Andes or Himalaya, ice and snow blanket the slopes of crustal blocks thrust upward by tectonic forces. Melting water seeps into every hairline crack or crevice, only to freeze again as temperatures drop. Water's peculiar property of expanding as it freezes acts as a molecular wedge, widening fractures and forcing rocks apart, crumbling mountains young and old. Once loosened, melting snow and ice carries the debris away, starting it on a liquid conveyor belt headed downhill for deposition in valleys or the oceans.

That process begins with rivulets trickling downhill, carrying mud and soil with them. Creeks and streams join to form rivers, scouring their banks and meandering their way seaward.

Rivers carry off about 20 billion tons of eroded material annually, enough to lower the continents by about an inch every thousand years. The sediment winds up in floodplains, in river deltas, on the offshore continental shelves, or on the deep ocean floor. So widespread and rapid is the process of erosion by running water that despite the great age of Earth, most of Earth's landscapes are no older than a few million years. Our planet's streams and rivers so thoroughly control the land that they erase traces of other processes, like tectonics, volcanism, and impact, that so dominate what we see on our planetary neighbors.

The makeup of the underlying bedrock and the slope of the land dictate how river valleys develop and mature. Young rivers in newly uplifted land cut steep-sided, V-shaped valleys, coursing down steep gradients and forming frequent rapids or waterfalls over resistant rock. Middle-aged streams follow a more winding course, fed by tributaries, and carving cliffs at the edges of wider valleys. Their slower water flow drops stones, pebbles, and sediments into the riverbed. Older rivers, now fed by an extensive drainage pattern, carry large volumes of water across nearly flat land in a series of wide bends called meanders. Upon reaching the sea, mature rivers finally give up their sediment to form extensive deltas of mud and silt. Periodic floods build natural dikes, or levees, and the shifting channel of a mature river leaves behind cut-off oxbow lakes.

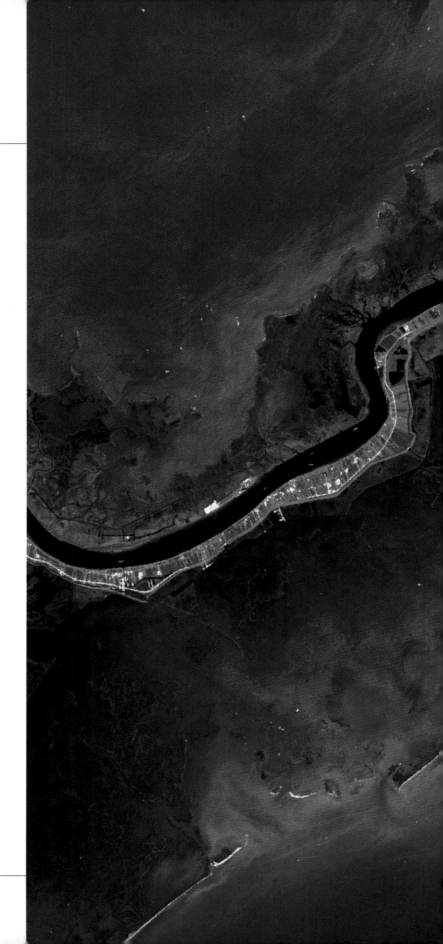

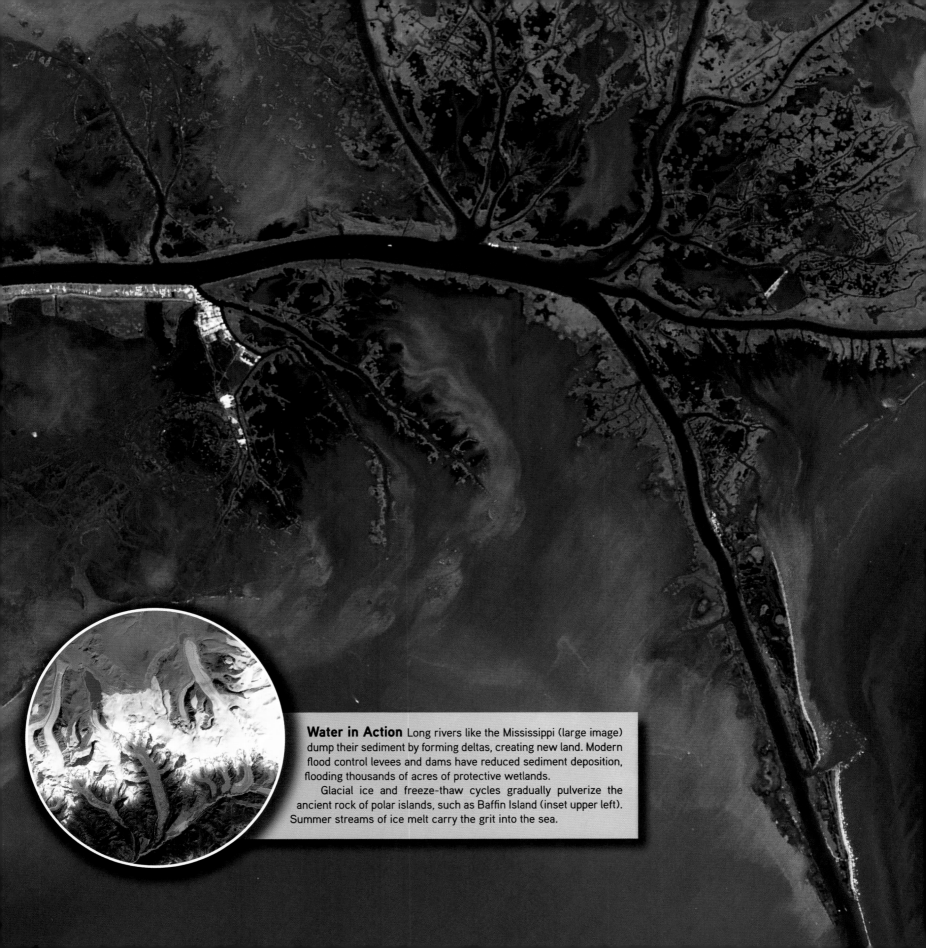

Water in Action Long rivers like the Mississippi (large image) dump their sediment by forming deltas, creating new land. Modern flood control levees and dams have reduced sediment deposition, flooding thousands of acres of protective wetlands.

Glacial ice and freeze-thaw cycles gradually pulverize the ancient rock of polar islands, such as Baffin Island (inset upper left). Summer streams of ice melt carry the grit into the sea.

Major Rivers of Earth

SEE ALSO: *Brahmaputra River* 102

BECAUSE RIVERS do most of the work of carrying the eroded remains of the continents seaward, it's worth taking a moment to examine their variety on our own world. The Brahmaputra and Ganges of India and Bangladesh are young rivers along their upper reaches, draining the towering Himalaya. Rapid erosion there—powerful monsoons and landslides—erodes two to four miles of rock from the mountains every million years. These two rivers discharge a sediment load at their common delta on the Bay of Bengal 20 times that of the Amazon.

Blasting through Hermit Rapid in the Grand Canyon, rafters are engulfed in the silt-laden waters of the Colorado River. The Colorado slices through its namesake plateau to carve a gorge nearly a mile deep.

the most recognizable landmarks to orbiting astronauts. Mature rivers like the Mississippi are constantly on the move, shifting into new channels and forming new deltas.

Floods are a hallmark of river systems on Earth and other planets observed across the solar system. On Earth, seasonal floods are common where annual rainy periods, spring snowmelt, or monsoons overwhelm the capacity of the drainage system. Floods cut new channels, scour sediments from upstream, and dump their loads to remake islands, build levees, and spread rich sediment

The lower Missouri River is a middle-aged stream, still carving its way through the badlands of the Dakotas and cutting cliffs into the surrounding plateaus. Draining the northern reaches of the Rockies, the Missouri furnishes most of the sediment the Mississippi carries after the two join at St. Louis. The combined Missouri-Mississippi Basin is the second largest in the world, after the Amazon.

The Mississippi and Amazon are examples of mature river systems, carrying vast amounts of water and sediment across nearly flat plains. Fed by runoff from the Andes and the tropical rains of the South American interior, the 6,570-mile-long Amazon disgorges into the Atlantic one-fifth of all the fresh water entering the oceans worldwide. The Mississippi-Missouri is nearly as long, but its temperate watershed sends only one-sixth the flow of the Amazon into the Gulf of Mexico. The sinuous channel of the lower Mississippi winds across a flat alluvial plain, marked by numerous oxbow lakes and traces of past floods. Its miles-deep sedimentary deposits are so massive that they have depressed the continental crust; its multichanneled delta, stretching far into the Gulf of Mexico, is one of

across adjoining tracts. The Mississippi floods about twice every decade; the 1993 flood after a wet spring runoff breached many of the river's protective levees and inundated the St. Louis area. On the Indian subcontinent, the floods of the monsoon-fed Ganges-Brahmaputra regularly put much of low-lying Bangladesh underwater. Regularly renewed and fertilized, the floodplains not only absorb destructive floodwaters but also are home to rich biological habitats.

Everywhere on Earth we see landscapes reworked by the action of running water: beaches, dry lake beds, flood-carved gullies and canyons, sand-and-gravel-filled braided river channels, floodplains, alluvial fans, flat-topped mesas, and the mighty courses of major rivers. These landforms, so common on Earth as to seemingly define a "normal" landscape, are direct results of the presence of a hydrologic cycle and the erosive power of water. When we first glimpsed these familiar landforms on other worlds, planetary geologists were both surprised and intrigued to find Earth was not the only planet shaped by this powerful agent. Evidence of water erosion on Mars was our first revelation.

Convergence Three rivers converge at St. Louis, Missouri: the Missouri (lower left), Mississippi (upper left), and Illinois (upper center). The largest flood ever recorded on the upper Mississippi River occurred after very heavy rains in 1993. The floods overwhelmed the Mississippi River Basin's elaborate system of dikes, dams, and levees (inset). Landsat 5 imaged the floodwaters near St. Louis in August 1993, slightly after peak water levels had been reached. Water appears dark blue, healthy vegetation is green, bare fields and freshly exposed soil are pink, and concrete is gray. The deep pink scars show where floodwaters have drawn back to reveal the scoured land. Many structures built within the natural floodplains were destroyed.

The Martian "Canals"

ARLY 20TH-CENTURY attempts by astronomer Percival Lowell to map the Martian surface led to exaggerated reports of "canals" on the red planet's surface. In 1965, Mariner 4 imaged a battered, desolate world, dashing hopes of finding practicing Martian engineers. In 1971 Mariner 9 found an unmistakable pattern of small stream networks that had once fed moving liquid downhill into successively larger canyons. The delicate branching pattern of these narrow, sinuous channels resembles nothing so much as Earth's natural river valley systems. Planetary geologists theorized that the likeliest explanation for the stream channels was that rainfall or spring water had once run across the surface, producing an erosion pattern typical of terrestrial stream valley networks. They knew that Mars's polar caps might harbor some water ice as well as frozen carbon dioxide (dry ice). Today's subsurface conditions would still permit liquid water to be present below the frigid surface.

The valley's narrow width (less than a mile), the intricate tributary system, and the fact that the valleys increase in size as they go downstream imply that they were cut slowly over time by running water, not by large floods. Under present climate conditions, liquid water cannot run in surface stream channels—it would freeze and quickly sublimate into the thin atmosphere. So the slow action of erosion represented by these valleys means Mars must once have possessed warmer temperatures and higher surface pressures. And because the valleys appear almost entirely in the older, cratered highlands of Mars, we can conclude that long ago the planet was warm enough to support the presence of liquid water on the surface.

This hospitable climate didn't last long. The numerous valley networks didn't lengthen into major river drainages or cut tributaries that captured the flow of adjacent drainages. This ancient era of Martian rainfall or groundwater runoff must have ended billions of years ago, with changing atmospheric conditions that eliminated the possibility of surface water.

The Huygens spacecraft captured this image of surprisingly Earthlike river channels and lake beds on Titan. A methane rain probably carved the snaking and branching riverbed (dark channel near top). The dark, smooth area (at bottom) appears to be a dry lake bed.

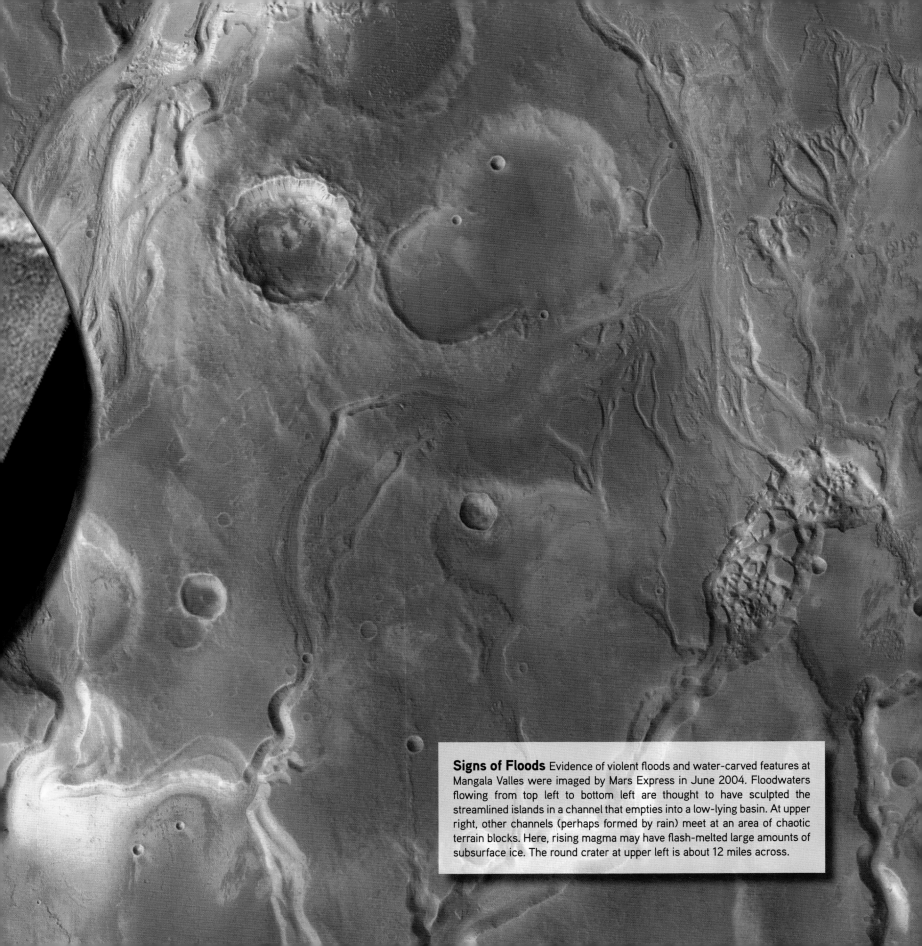

Signs of Floods Evidence of violent floods and water-carved features at Mangala Valles were imaged by Mars Express in June 2004. Floodwaters flowing from top left to bottom left are thought to have sculpted the streamlined islands in a channel that empties into a low-lying basin. At upper right, other channels (perhaps formed by rain) meet at an area of chaotic terrain blocks. Here, rising magma may have flash-melted large amounts of subsurface ice. The round crater at upper left is about 12 miles across.

SEE ALSO: *Life on Mars* 178

Carved by Floods

THE SOLAR SYSTEM'S grandest canyon—named Valles Marineris after its spacecraft discoverer, Mariner 9—is a tectonic rift valley that stretches across Mars's equatorial region. But portions of Valles Marineris, as well as dozens of large channels across the planet, exhibit flow features that must have been carved by floods. The channels rise abruptly from depressions, surrounded by inward-facing walls containing jumbled blocks of bedrock. From this "chaotic" terrain, some channels extend thousands of miles from their source regions until they merge with low-lying plains high in the northern hemisphere. The Martian channels possess few tributaries and appear to emerge at full flood from the strange terrain. Their courses are marked by dramatic flow lines, scour marks, and teardrop-shape islands, just like those in sandy-bottomed terrestrial streams. The only difference is the huge volume of water these flood channels carried—when full, the Ares Vallis channel carried a peak flow more than a thousand times that of the Mississippi River.

What caused these immense flood outbursts? Geologists noted the similarity of the Martian channels to outsized flood features in eastern Washington State. But unlike the Channeled Scablands region of the Pacific Northwest, formed by a series of collapses of glacial ice dams during the last ice age, the Martian flood channels were probably created when subsurface groundwater, held under pressure beneath a frozen permafrost layer, breached this barrier, perhaps through faulting or an asteroid impact. The tremendous rush of pressurized water to the surface collapsed and eroded the overlying rock and sediment, leaving wildly jumbled blocks behind, carving immense scour marks and streamlined islands. Rock barriers along the way might have created temporary lakes; when these failed, a series of cascading, catastrophic floods occurred. Lake deposits seen in the precipitous canyon walls of Valles Marineris suggest that water briefly ponded in sections of the canyon behind natural dams, then finally broke through and scoured outburst channels within the valley itself.

Resembling the braided islands of the Betsiboka River Delta in Madagascar (right), Ares Vallis is one of many flood channels formed when water burst from chaotic terrain in Mars's southern highlands (below). These "islands," some more than ten miles long, were deposited by floods. Floodwaters also carved the parallel grooves in the channel floor.

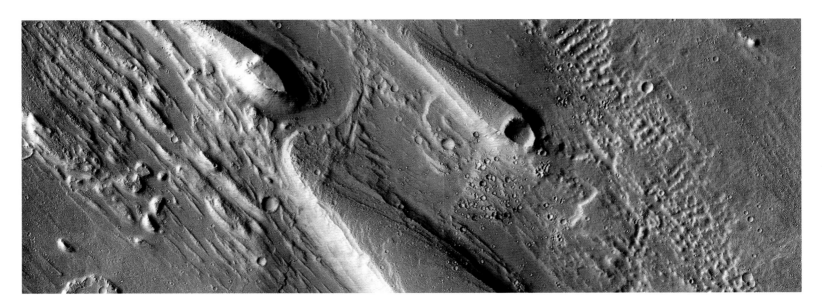

Ancient Lakes

SEE ALSO: *Life on Mars* 178

WHAT HAPPENED TO THESE floodwaters on Mars once they reached the Martian lowlands? Orbiting spacecraft have found many instances on Mars of layered lake sediments and terraces seemingly lapped by the waters of broad lakes, perhaps even shallow seas.

Most of the layered rocks on Mars—which seem similar to those formed predominantly on Earth by sediments deposited underwater—are within large craters that once may have pooled floodwaters for extended periods. The pancakelike layers, stacked one atop another, are typical of terrestrial sediments lain down in lakes or shallow arms of the sea. Scientists can't tell from the images where the sediments came from: washed in by floods, precipitated chemically out of solution, or blown in as dust by Martian winds. The material in the layers is eroding at cliff faces today into granular soils and dust, not large boulders, so the layers may consist of fine-grained deposits of clay or silt. The waters in Hebes Chasma (left) appear to have persisted after the early period of stream valley formation, extending the time in which the planet hosted wet, hospitable conditions for life. Geologists and astrobiologists hope investigations of these Martian sediments will turn up a detailed record of Mars's climate and its suitability for life.

A man runs across the dry salt flats of the Salar de Atacama in Chile (large image at right), the layered sediments of an ancient lake. Lakefloor sediments may also be the source of sediments eroding from the face of the mesa (above) in Hebes Chasma, a nearly five-mile-deep canyon on Mars. The stacked rock layers might have been laid down by floodwaters pooled in the gorge.

GREAT SALT LAKE

Utah's Great Salt Lake is the remnant of the much larger Lake Bonneville, which existed between 32,000 and 14,000 years ago. As the cool wet climate changed to a drier one, evaporation exceeded the flow into the lake from runoff and rainfall, shrinking the lake to its present shoreline and making the once fresh waters briny; about two million tons of minerals and salts wash into Great Salt Lake annually. With more than 4.5 billion tons of salts dissolved in the lake, the average salinity of the south arm of the lake is about 13 percent, enough for swimmers to float easily in the dense liquid. The 1,600-square-mile lake grows and shrinks with local climate variations. Because geologic faulting has caused the basin to slowly subside, about 12,000 feet of lake sediments have accumulated beneath the lake's floor. If uplifted, exposed, and eroded, those sediment layers would offer a valuable climate record—one that would help scientists understand not only Earth's past but possibly unravel the origins of the layers of sediment in Mars's Hebes Chasma.

SEE ALSO: *Martian Canals* 110

Water on Mars Today

THE MARS GLOBAL SURVEYOR CAMERA, operational in orbit from 1997 to 2006, repeatedly examined geologically youthful gullies in the walls of Martian craters. The original photos hinted that erosion had formed gullies very recently in Martian history, but the images could not prove that liquid water was the agent. Then in November 2006, researchers comparing early and recent images discovered that those gullies had changed in the past five years. The simplest explanation was that fresh material was washed down the gullies by liquid water. The water might have emerged from a subsurface layer in the wall of the crater, flowing so vigorously that it carried debris down to the crater floor before evaporating. It's not quite a smoking gun (or dripping hose), but the gully images are the best proof yet that water is still reworking the face of the red planet. These sites are prime candidates for future investigations by landers, rovers, and human explorers.

Although the Mars exploration rovers have not seen any evidence of water-driven erosion in their four-year-long traverses across the Martian plains, they nevertheless found proof that water was once present. The rover Opportunity, sampling rock outcrops, found the mineral jarosite, which typically forms when rocks are saturated in sulfate-rich standing water. Opportunity also found thousands of tiny spherules, nicknamed "blueberries," and rock cavities typical of mineral formation and dissolution in a long-lasting aquifer. It appears the bedrock in Meridiani Planum, Opportunity's landing site, was once drenched in salty water, either by an aquifer or as the floor of a shallow sea, like Utah's Great Salt Lake. Today, briny groundwater may persist below the surface, escaping to the surface as seeps or carving small, fresh gullies. And water is still present in the polar caps; the Mars Express satellite detected enough ice there in 2007 to submerge the entire surface in more than 30 feet of water!

Applying our knowledge of Earth's water-dominated geology and mineralogy has provided us the insight to unravel part of Martian history. From a youthful period of massive floods to today's mere hints of liquid water, we now know that the story of Mars, and especially its potential as a harbor for life, is, like Earth's, carried along by the flow of water.

Images of gullies found obvious geological changes on the interior slopes of a Martian crater between 2001 and 2005. A mini-gully washer of briny water, bursting from an underground source, seemed a strong candidate to explain the bright deposits; however, a 2007 analysis found that a simple rockslide of dry sediments could explain the visible changes.

gully as it appeared in 2001

150 m

gully as it appeared in 2005

Finding Blueberries In the Grand Staircase-Escalante National Monument in southern Utah, marble-size pebbles known as hematite concretions litter the Navajo sandstone surface. These pebbles accumulated after wind and water eroded the softer surrounding sandstone away. In a much harsher but strangely familiar desert environment, Mars rover Opportunity's microscopic imager in 2005 captured the image of so-called blueberries, plentiful mineral concretions found embedded in bedrock layers and scattered on rock surfaces across the Martian landscape. They are graphic evidence that the bedrock in Meridiani Planum was once saturated in mineral-rich water, providing conditions in which the spherical nodules could grow.

Cold Lakes

BEYOND MARS, solar system surfaces are simply too cold for liquid water. On the ice-bound satellites of Jupiter, for example, subsurface water may rarely escape through a crustal fissure, flooding and repaving local terrain in ice, but it can't stay liquid long enough to flow and shape the surface. At Saturn's largest moon, Titan, the Cassini spacecraft has finally been able to reveal details of the smog-shrouded surface, and the results have been startling.

When Cassini's Huygens probe streaked into Titan's upper atmosphere in 2005 and popped out a parachute for a first ever descent through the thick nitrogen atmosphere, some scientists thought it might splash down into an ocean of hydrocarbons; others expected a cratered, icy surface. But the probe revealed instead a complex landscape of sinuous, branching valleys, apparently carved by some kind of flowing liquid.

Water? No. Titan's surface at minus 289°F is far too cold for that. But it is warm enough for liquid methane—natural gas on Earth—to fall from the atmosphere as rain, and pool on the water-ice surface. The images from Huygens showed a ridge near its landing site cut by a dozen dark lanes, or channels. Elevated terrain nearby is laced by a complex channel network, with tributaries feeding into larger streams, and eventually into a dark, smooth valley floor. Huygens landed on a solid surface of soil and pebbles, but its instruments detected a whiff of methane, perhaps expelled from saturated sand. The runoff channels imply that liquid from precipitation (likely as methane "rain") or springs periodically carves the icy landscape, just as water does on Earth and once did on Mars. In fact, the Titan channels may be analogous to those in arid regions on Earth, where occasional, perhaps seasonal, rainstorms alter the surface.

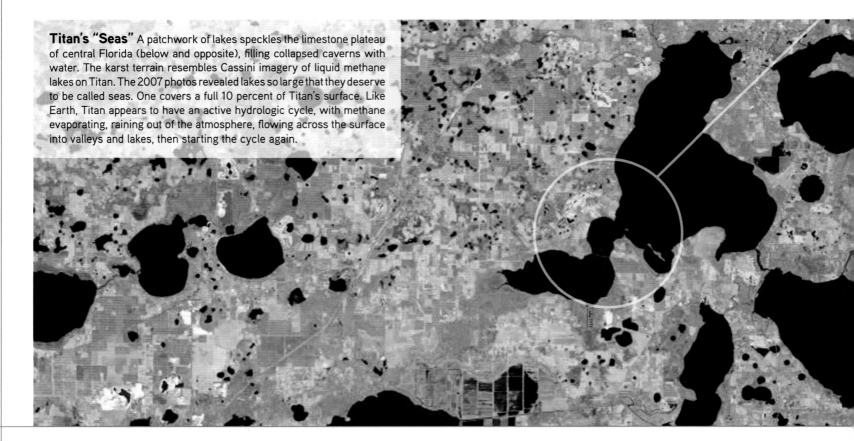

Titan's "Seas" A patchwork of lakes speckles the limestone plateau of central Florida (below and opposite), filling collapsed caverns with water. The karst terrain resembles Cassini imagery of liquid methane lakes on Titan. The 2007 photos revealed lakes so large that they deserve to be called seas. One covers a full 10 percent of Titan's surface. Like Earth, Titan appears to have an active hydrologic cycle, with methane evaporating, raining out of the atmosphere, flowing across the surface into valleys and lakes, then starting the cycle again.

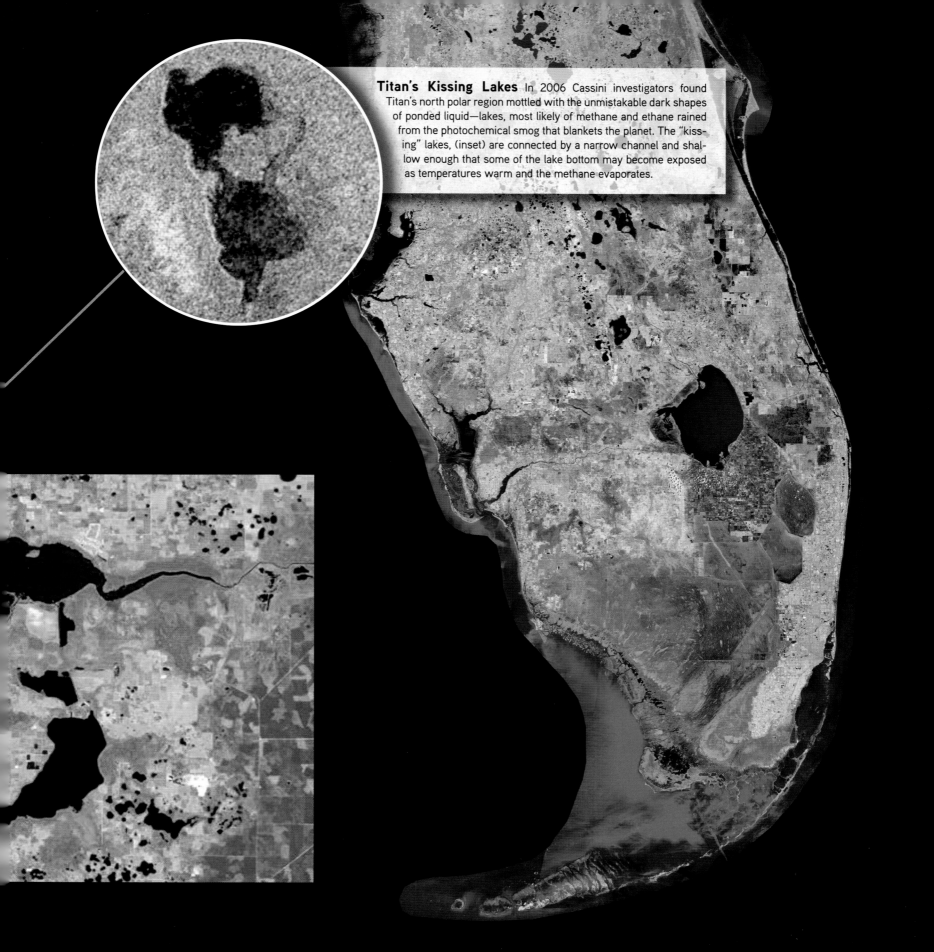

Titan's Kissing Lakes In 2006 Cassini investigators found Titan's north polar region mottled with the unmistakable dark shapes of ponded liquid—lakes, most likely of methane and ethane rained from the photochemical smog that blankets the planet. The "kissing" lakes, (inset) are connected by a narrow channel and shallow enough that some of the lake bottom may become exposed as temperatures warm and the methane evaporates.

SCORCHING VALLEYS

One might think Titan's frigid methane lakes are the most exotic examples of surface fluids in the solar system, but geologists find, at the other end of the temperature spectrum, another rare but powerful fluid force shaping planetary landscapes. Volcanoes erupt molten rock. As this lava seeks a lower level under the force of gravity, its immense heat can reshape the rock over which it flows into dramatic channels.

I hiked such lava channels during my shuttle geology training on Hawaii's Kilauea volcano, and with our spaceborne radar we scanned other shield volcanoes around the world to relate these features to similar ones discovered on the rocky planets. On the moon, several "sinuous rilles" are probably lava tubes or channels that fed the vast basaltic lava lakes that once pooled in the maria (dark lunar "seas"). The Apollo 15 astronauts landed next to a huge rille, Hadley. They stood at the canyon's rim and photographed the basalt layers exposed on the opposite wall, but the slopes were too dangerously steep to enable them to descend and sample them. Almost a mile wide and a thousand feet deep today, Hadley's lava torrent probably melted its way through several older lava beds when it formed 3.3 billion years ago. Some portions of the lava channel are still roofed over, forming a lava cave, or tube. The greater volume of lunar basalt eruptions, combined with the moon's lower gravity, created spectacular lava-cut valleys that dwarf those of Earth.

On the moon, lava stopped flowing over two billion years ago, but Venus's surface might still be shaped by molten flows. More than 200 lava channels have shown up in radar images of Venus. Most are found amid larger outpourings of lava or on gentle volcanic slopes, but about 50 run far from existing volcanic structures. Some show older, cut-off channels or meanders, like mature river valleys on Earth, suggesting they might have been formed by successive lava flows. One channel is longer than the Nile River—and at 4,200 miles long, is the longest lava channel anywhere in the solar system.   —Tom Jones

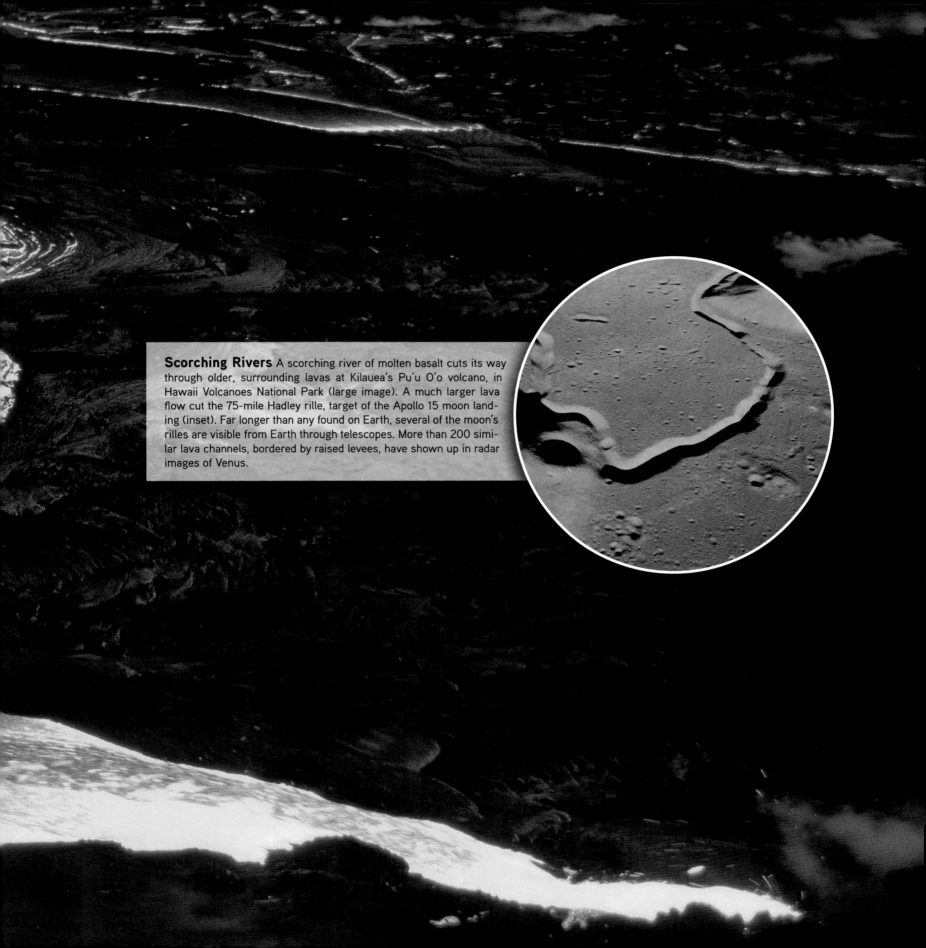

Scorching Rivers A scorching river of molten basalt cuts its way through older, surrounding lavas at Kilauea's Pu'u O'o volcano, in Hawaii Volcanoes National Park (large image). A much larger lava flow cut the 75-mile Hadley rille, target of the Apollo 15 moon landing (inset). Far longer than any found on Earth, several of the moon's rilles are visible from Earth through telescopes. More than 200 similar lava channels, bordered by raised levees, have shown up in radar images of Venus.

Watery Worlds

SEE ALSO: *Europa's Watery Secrets* 180

ON VENUS, THE MOON, AND SATURN'S IO, the erosive fluid at work has been not water but lava. But Venus's lava channels, like those on the moon, have reworked only a tiny fraction of the surface. Molten rock cannot match the ubiquitous power of liquid water. On the planets, lava flows from only a few sources, for brief periods of time, while the hydrologic cycle on Earth has shaped our planet's continents for billions of years.

We see evidence for fluids on a variety of intriguing bodies in the solar system. We know that on Saturn's moon Enceladus liquid water or ice crystals are jetting from geysers along surface cracks in the icy crust. Jupiter's satellite Europa probably has a substantial water ocean beneath its solid ice crust. Titan probably harbors a similar water layer, far beneath its its frozen surface, kept liquid by Saturn's tidal flexing. Even the largest asteroid, Ceres, might have water-saturated rocks beneath its pitted surface. But water or other volatile compounds, like methane, require a substantial atmosphere to keep them in a liquid state at the surface long enough to cause erosion. Enceladus, Europa, and Ceres all lack atmospheres, so any water that might appear on the surface either instantly freezes or sublimes quickly into space. It cannot rework its host world.

Water's presence seems to be a prerequisite for the existence of life. If Earth were not a water planet, no one would be here to note how widely that fluid has shaped our surroundings. Earthquake-caused tsunamis scour our populated shorelines, and rivers deposit millions of tons of silt to build huge river deltas. Water laid down Earth's extensive beds of sedimentary rock; today it nourishes our farmlands and cities. Its powerful floods destroy lives and alter our landscape overnight. From the depths of the Grand Canyon to the stark and ancient islands of the Martian flood channels, water is the force we humans recognize first. For better or worse, we live on a world where its work continues.

Earth's watery surface is serene—at peace—in this STS-80 space shuttle view of the Pacific, west of Hawaii.

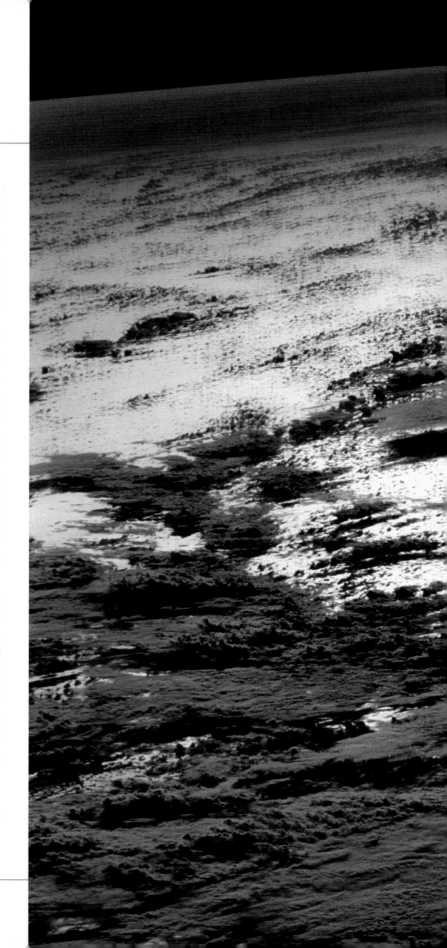

AN UNSTOPPABLE FORCE

An astronaut's eye takes in the sweep of water's beauty and power: sunglint from ocean, lakes, and rivers; the fingerprint whorls of ocean swells, marching relentlessly between continents; the exquisite swirl of clouds above a hurricane's raging storm surge. Water driven by wind, tumbling toward the sea, or heaved by tectonic motion is a nearly unstoppable force. In December 2004, a magnitude 9 earthquake shifted a huge block of seafloor off the island of Indonesia, creating a massive tsunami. The surge of seawater onshore, and its subsequent retreat, wiped out entire villages, like this one on the coast of Sumatra (above). More than 225,000 people died. On Mars, cataclysmic floods of water once carved canyons, turned impact craters into lakes, and created ephemeral seas. Much of our focus in exploring Mars today is a search for the source of that water, and where it might reside today. Water frost at the south polar cap, if melted, is sufficient to cover the planet to a depth of about 36 feet. Many craters in mid- and high latitudes display ejecta blankets—layers of impact debris—which suggest the impact threw out a fluid slurry of rock and mud. Subsurface permafrost might have created such characteristic craters, and we suspect that frozen soil may still harbor significant reservoirs of water. The Mars Phoenix lander, which touched down near the planet's north pole on May 25, 2008, will measure the water content of polar soils and use its robot arm to dig for permafrost.

FROZEN LANDSCAPES

Ash from the volcano Mount Belinda darkens an iceberg in the South Sandwich Islands, near Antarctica. While we do not find penguins on other worlds, we do find frozen landscapes.

SEE ALSO: *The Future of Ice* 144

Our Icy Past

TWENTY THOUSAND years ago, the Earth viewed from space would not have looked like our familiar blue planet with its ice-coated poles. Instead, an orbital observer would have seen sheets of ice extending from the North Pole to the Great Lakes, and covering most of northern Europe. Over Earth's long history, ice has periodically advanced and retreated across the surface of the planet, changing global climate and causing extinction of species. Our last ice age ended about 10,000 years ago. Today, ice sheets are confined to the Poles and Greenland, with glaciers present, though many of them are melting, on all of the continents and Iceland. About 90 percent of Earth's fresh water is stored in ice caps and glaciers; if it all melted, Earth's sea level would rise as much as 229 feet.

Ice is present throughout the solar system. It has been detected on Mercury and the moon, and on the icy satellites of Jupiter, Saturn, Uranus, and Neptune. At present though, only Mars has clearly visible polar caps of ice. Images of the Martian surface also show evidence of the past action of glaciers on the surface. More controversially, it has been suggested that Mars has a frozen sea.

Earth's polar regions are covered by ice sheets or ice fields—piles of ice that accumulated by precipitation or condensation, whose shape is not controlled by hills or valleys. Snow that falls on the ice cap or sheet is compacted over time, forming layers of ice. The ice at the bottom can be very old; some ice cores from Antarctica contain ice from more than 750,000 years ago. Some of this ancient ice contains trapped bubbles that allow scientists to study how the atmosphere of Earth has changed over time. The weight of these giant ice sheets on Earth's crust actually causes the crust to bend downward, or subside. The center of Greenland

During ice ages, species had to adapt to the much greater extent of ice and colder temperatures. Mammoths (above) thrived from about 4.8 million years ago to about 10,000 years ago, dying out at the end of the last ice age.

has been depressed below sea level by the weight of the overlying ice sheet, almost two miles thick in places. As an ice sheet melts, relieving the pressure, the resilient, plastic-like mantle under the crust will rise back up again, a process called isostatic rebound.

The polar regions of any planet receive slanting, indirect light from the sun and thus never reach the warm temperatures experienced by equatorial regions. The tilt of Earth's axis produces darkness for much of the polar winter, lowering temperatures further. Prolonged freezing conditions enable permanent caps of ice to reside at both poles, with the southern ice sheet lying mainly over the continent of Antarctica, and the north polar cap covering the Artic Ocean. The extent of sea ice at the margins of the caps shrinks in summer and grows in winter. Earth's changing climate controls the long-term shape and coverage of the ice caps, which in past eras have nearly disappeared or spread far into the mid-latitudes. Portions or all of both caps are currently shrinking in overall size due to global warming.

Mars has ice caps at both its poles. The Martian caps are made of water and carbon dioxide ice, with the north polar cap composed primarily of water ice and the southern cap around 90 percent water ice. The ice caps are surrounded by polar layered deposits—piles of rock and dust mixed with ice. The two caps are similar in size, up to 2 miles thick and more than 620 miles across. In the winter, both polar caps accumulate several feet of carbon dioxide ice, deposited out of the atmosphere, which then sublimes away in the summer, going from a solid directly to a gas. It is never warm enough for the water ice in the caps on Mars to melt; the low atmospheric pressure causes the ice to sublime instead.

EXPLORING THE POLES

Exploration of the Poles did not occur with any success until the 20th century, due to extreme difficulties in traveling long distances over ice. Temperatures at the South Pole in the summer only reach about minus 12°F; summer temperatures at the North Pole are much warmer at about 32°F. Roald Amundsen was the first to lead a successful expedition to the South Pole, arriving on December 14, 1911. He reached the Pole one month before the British expedition led by Robert Falcon Scott. While Amundsen made it safely home, Scott and his four companions, who reached the Pole, died of exposure and scurvy on their return trip. In 1914, Ernest Shackleton attempted to cross Antarctica via the South Pole. Shackleton heroically rescued his crew after his ship, the *Endurance* (inset), was trapped in pack ice and sank. It was not until 1958 that the next overland expedition reached the South Pole. North Polar exploration was no less difficult. Two Norwegian explorers, Fridtjof Nansen and Fredrik Johansen, came within a few degrees of latitude of the Pole in 1895. In April 1908, a group led by American Robert Peary were the first to actually reach the Pole, although this claim remains controversial.

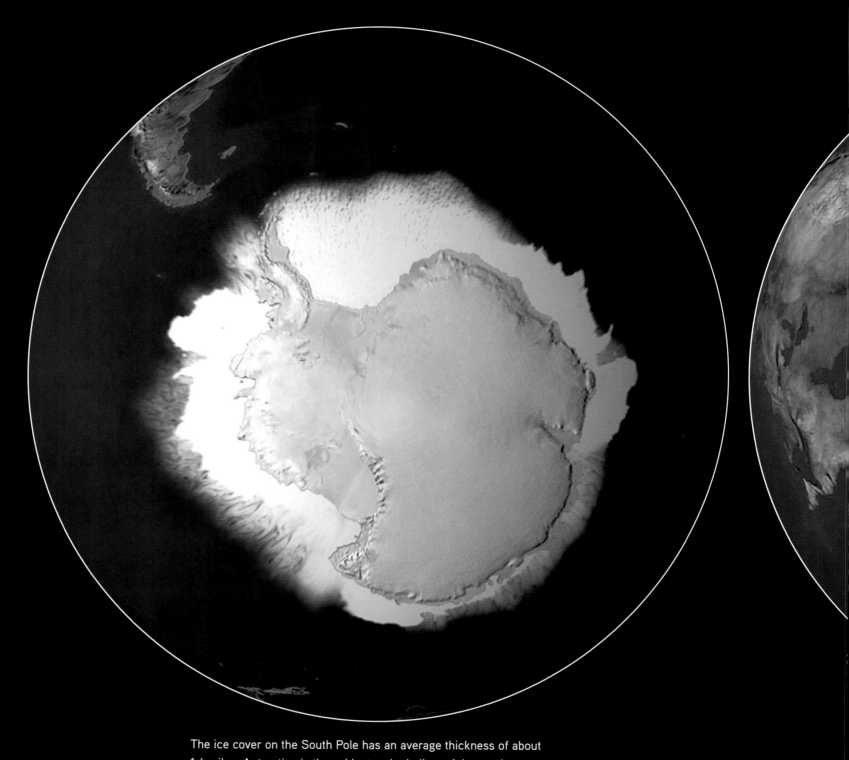

The ice cover on the South Pole has an average thickness of about 1.4 miles. Antarctica is the coldest and windiest of the continents.

At its winter maximum in 2008, ice covered more than 15.87 million square miles around the North Pole.

The Poles on Earth are located where the imaginary line about which Earth rotates cuts through the planet. The rotational pole does not stay in one place—it wobbles over a distance of a few feet every year. To confuse matters further, the rotational poles are not the locations of the magnetic poles. The magnetic poles, like Earth's entire magnetic field, shift constantly; they can move as much as nine miles in a year. The magnetic poles are defined as the point where the flux of Earth's magnetic field points downward, so the location changes as the magnetic field varies. Even the direction or polarity of the magnetic field can change— magnetic north and south have switched multiple times over Earth's history for reasons that are not well understood. The last switch in field was about 780,000 years ago. The strength of the magnetic field has been declining over the last 150 years, with a recent increase in the rate of decline. It is thought that the field will decline before a reversal, then either completely or nearly vanish, and then restrengthen in the opposite sense—compass needles that used to point north would then point south.

Earth's magnetic field protects us from the wind of energetic charged particles streaming out from the sun. A full reversal of the field would be very disruptive to communication and navigation, but there is no link between species extinctions and past field reversals. Magnetism is related to the composition and interior structure of planets, and is present when a portion or all of a planet's core is fluid. The motion of the interior fluid, generated by the planet's rotation, creates a magnetic field. Mars and Venus have no detectable magnetic fields, which indicates that their cores may be solid. However, magnetic signatures in surface rocks on Mars indicate that Mars had a magnetic field in the past. Mercury does have a weak magnetic field, and its very large core may still have a fluid interior. The solar system's gas giants—Jupiter, Saturn, Uranus, and Neptune—all have magnetic fields that are much stronger than Earth's. Jupiter's magnetic field is unique; scientists have observed a current passing between the gas giant and its moon Io.

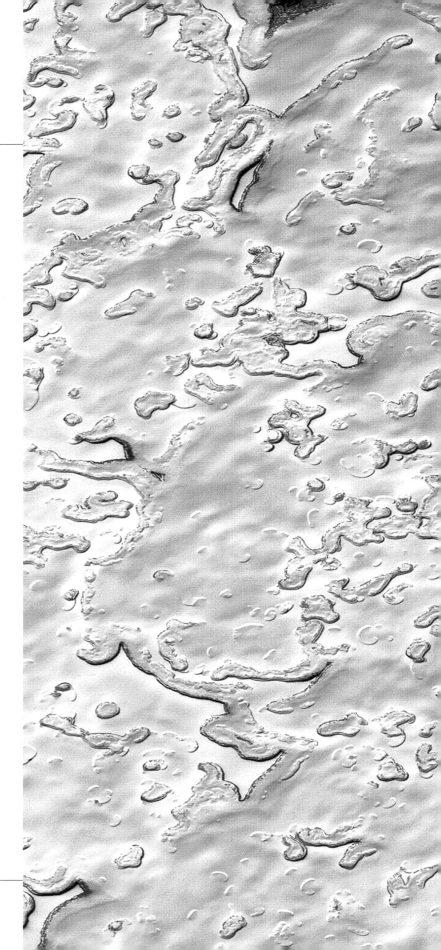

Polar Ice on Mars

SEE ALSO: *Ice Volcanoes* 140

THE POLAR ICE ON MARS contains layers of sand and dust that can be seen in images of troughs and canyons that slice deep into the caps. These layers, as little as a few feet thick, record changes in the climate of Mars—like those that caused the ice ages on Earth—probably due to slight changes in the planet's orbit. Some of the layers at both poles are wrinkled and deformed, indicating movement of the ice. Both caps have long, curving canyons, a half mile deep and as much as 60 miles wide, cutting into them in a spiral pattern. The largest such valley at the north pole is called Chasma Boreale, at the south pole, Chasma Australe.

The surfaces of the caps show evidence for sublimation or ice loss, wind erosion and collapse of the ice; and they exhibit a few craters. All of this indicates that the upper surfaces of the caps are relatively young, hundreds of thousands to tens of millions of years old. They

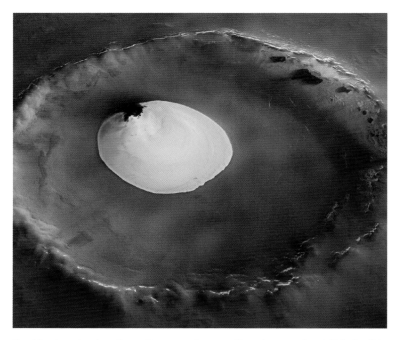

Residual water ice from the receding north polar cap is visible in this image on the floor of Vastitas Borealis crater on Mars.

Changing Patterns The surface of Mars's south polar cap (at right) changes seasonally in extent—in winter this scene would be covered by frost. Sublimation or loss of ice and sculpting by the wind results in beautiful patterns of ice and sediment, as seen in the image from Mars Global Surveyor (large image) taken in 2001. The Martian polar caps changed in extent over time, leaving behind many characteristic glacial features. By studying the nature and geographic extent of polar landforms, scientists can study how the climate of Mars has changed over time.

have clearly moved over time, leaving behind many characteristic landforms that are reminiscent of glacial features on Earth. Deposits of ice can be seen within some craters, left behind when the cap receded. The Mars Express sounding radar has estimated the volume of the south polar cap to be equal to a global ocean about 36 feet deep.

While the heavily cratered surfaces of the moon and Mercury seem unlikely places to go in search of ice caps, evidence of water ice has been detected at their poles. On the moon, high concentrations of hydrogen have been measured at the poles, suggesting that there is ice present at or near the surface. The temperature in direct sunlight would be enough to evaporate water ice, but polar impact craters with areas in permanent shadow could preserve ice over long time periods. Water ice would be of use to astronauts living and working on the moon to support a lunar outpost and possibly make rocket fuel from the hydrogen and oxygen. The composition of these ices may also provide information on comets, a likely source of the ice.

Avalanche! Avalanches occur when materials on slopes become unstable, whether that material is snow, as in the avalanche on Earth in the Caucasus Mountains in Russia (inset), or of dirt and ice on Mars. The High Resolution Imaging Experiment (HiRISE) on NASA's Mars Reconnaissance Orbiter caught an avalanche in action (large image) on February 19, 2008. The full image shows an area 3.7 miles wide and more than 35 miles long, near the north pole at 84 degrees north latitude. The reddish layers of the steep, 2,300-foot-tall slope are made up of water ice and dust. It is not clear what caused the ice and dust to avalanche down the slope. Further observations of the area will allow scientists to estimate the amount of ice in the avalanche deposit.

Rivers of Ice: *Glaciers*

EARTH IS HOME TO more than 67,000 glaciers. They are found on every continent—including Africa (on volcanoes Mount Kenya and Mount Kilimanjaro, and the Ruwenzori Massif) and Australia (on Heard Island, an island that is part of Australia). Glaciers vary greatly in size, covering small valleys or streaming across entire regions. Today glaciers cover about 10 percent of the land area of Earth—in the last ice age they covered more than 30 percent. The longest glacier in North America is the Bering Glacier in Alaska, 126 miles long. Most of Earth's glaciers are retreating, or becoming smaller, as the climate warms. Mount Kilimanjaro in Kenya may soon lose its famous snows, and Glacier National Park in Montana may be free of glaciers by 2030.

Unlike the ice sheets at the Poles, glaciers are smaller "rivers" of ice that are in motion, flowing down slopes under the force of gravity and usually confined in mountain valleys. Like ice fields, they collect snow and ice during winter, which become compacted into dense layers of ice. The top layers of ice are rigid and fracture into cracks called crevasses,

In 1991, a well-preserved body from 3300 B.C. was found in a glacier in the Ötztal Alps in Italy. Nicknamed Utzi, the "ice man" was found to have bled to death at age 45 after being hit in the back by an arrow, although he had also suffered a severe head wound. The artifacts and body are amazingly well preserved, from his arrows (above) to his clothing and the contents of his stomach.

while the lower layers flow more plastically downhill, riding on a base of meltwater. As the glacial ice moves downhill, it erodes the underlying rock, scouring out rocks and boulders that then continue to grind the surface as they are carried along with the glacier. Glaciers move at various speeds depending on the slope, thickness, temperature, and type of surface they rest on, from as little as 6.5 feet a year to as much as 5 miles a year. Some glaciers alternate between little to no movement and more rapid movement—up to dozens of feet a day. A sudden rapid downslope movement of a glacier is called a surge, probably caused by changes in the interior and base structure of the glacier.

As a glacier melts, it leaves behind a characteristic landscape. A glacier carves mountaintops into peaks called horns and arêtes that sit above bowl-shaped depressions called cirques; after a glacier melts completely, it can leave behind small lakes called tarns. As the glacier moves downhill, the heavy ice carves a U-shaped valley—rivers and streams carve V-shaped valleys. Rocks and dirt eroded and carried by the glacier, called glacial till, form ridges of material, or moraines, at the edges and ends of the melting glacier. For example, a terminal moraine (material left at the end of a glacier) can be found in the center of Long Island in New York—marked by mounds of till. This moraine formed about 21,000 years ago. Much of the landscape of present-day Scotland and Norway has been shaped by glaciers, with deep glacial lakes and U-shaped valleys.

Long, S-shaped mounds of till called eskers and more streamlined hills called drumlins are left behind after a glacier melts. Large boulders called glacial erratics also can be left behind as a glacier melts. These are commonly found in the midwestern United States, but also in such unlikely places as the top of Mauna Kea, in the Hawaiian Islands. Even this tropical volcano once was capped by glaciers. Ice blocks left behind can eventually melt to form depressions called kettles. Most of these kettles are now lakes, as seen in Minnesota and much of Canada. Scientists look for all of these distinctive glacial features on other planetary bodies.

FROZEN **LANDSCAPES**

Remains of Glaciers It is hard to imagine the beautiful valleys of Yosemite National Park in California (inset) covered in ice, but their distinctive U-shaped profiles tell another story. At Bear Glacier in Alaska (large image) imaged by the IKONOS satellite, ice breaking off, or calving, at the end of the glacier provides evidence that the glacier is moving and carving the landscape, and, in this case, actually melting away. The lake at the end of the glacier is a blue-green color typical of glacial lakes—a result of fine sediments from the glacier that are suspended in the water. Cracks in the surface of the glacier, or crevasses, can be extremely deep and pose a hazard to explorers.

Layers of History Ice sheets and glaciers grow and shrink with the seasons, accumulating ice in the winter, and losing it in the summer. This layering can be seen in the Franz Joseph Glacier in New Zealand (large image), as ice falling into the Waiho River reveals the internal structure of the glacier. On Mars (inset), this sloping edge of the south polar cap reveals the layers of ice and dust that have accumulated to form the ice caps. On Earth, cores into glaciers and ice sheets are used to study changes in the climate over time. Ash and other particles in the ice can be used to track volcanic eruptions and industrial pollution. Bubbles trap atmospheric gases allowing past atmospheric gas concentrations to be measured. The thickness of layers reveals the seasonal variations in precipitation over time.

Ice on Mars: *Further Evidence*

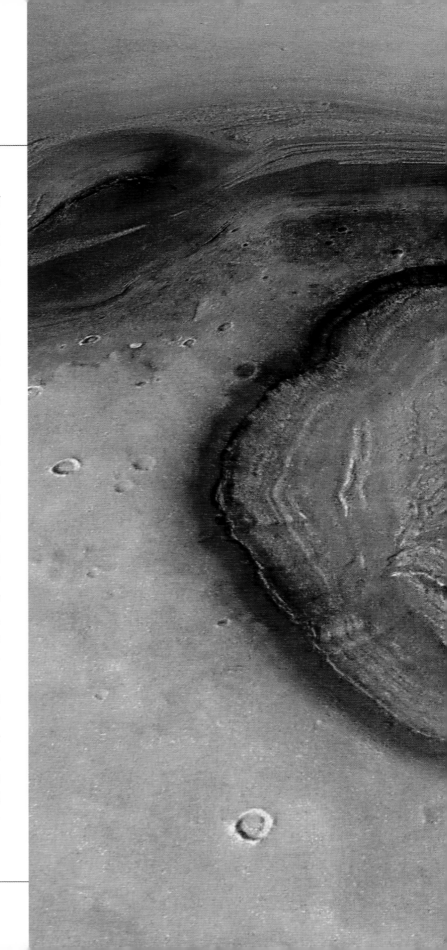

W E KNOW THAT WATER once flowed across the surface of Mars, carving great river channels. But what about frozen rivers? The slopes of the Arsia Mons volcano, a fan-shaped deposit about 70,000 square miles in size, may hold the remnants of a glacier. Ridges interpreted to be moraines surround the deposit, which may still contain ice mixed in with the rock and dust. Scientists studying the deposit using orbital photography could see no evidence for erosion caused by meltwater, suggesting that the glacier disappeared through sublimation, going directly from its solid state to vapor, rather than melting. Sublimation also takes place on some glaciers in Antarctica. Imagery has turned up similar glacial-type deposits on the flanks of other volcanoes on Mars as well.

In many regions of Mars, planetary geologists have spotted what appear to be glacial landforms including eskers, moraines, kettles, cirques, and arêtes, in patterns that suggest that glaciers may have covered much of the Martian southern and northern hemispheres at various times in the past. Ridges in the northern plains of Mars, called thumbprint terrain, resemble moraines and eskers and suggest that the area was once covered by a large glacier. Periods of glaciation might have occurred as recently as a few million years ago. Features associated with large-scale melting of glaciers are scarce, suggesting that they largely sublimed away. Cold-based glaciers, such as those in the frigid, dry valleys of Antarctica, often behave this way. Most Earth glaciers are warm-based—with bases at the melting point of water, due to the friction of their motion and heat coming from the ground beneath.

Signs of past glaciers on Mars are seen not only near the poles, but also much closer to the Martian equator, especially near the tops of volcanoes. Models suggest that glaciers could have formed near the equator if the tilt of the planet's rotational axis was larger in the past, a configuration which is known to have occurred as recently as 5.5 million years ago. This more pronounced tilt would cause more extreme seasons and put more water in the atmosphere, which could have resulted in the formation of glaciers at the tops of volcanoes like Arsia Mons.

Glaciers on Mars The large image on this page, taken by the High Resolution Stereo Camera (HRSC) on board the European Space Agency's Mars Express, shows a perspective view of a glacial feature located in Deuteronilus Mensae on the northern plains of Mars. It is remarkably similar to the Taylor Glacier in Antarctica (inset). On Mars, the ice of the glacier's surface has probably sublimated away, but ice is likely trapped in the sediments left behind.

Ice Volcanoes

ADDITIONAL EVIDENCE FOR extensive ice cover in Mars's past comes from Martian volcanoes. On Earth, volcanoes that erupt underneath sheets of ice have distinctive shapes. Called tuyas, these are flat-topped, steep-sided volcanoes that initially erupt under ice, but then eventually melt their way through. Tuyas are found in western Canada, but also in Iceland, where there are volcanoes currently erupting beneath the Vatnajökull glacier, the largest in Europe. Tuya-shaped volcanoes have also been identified on Mars, in the Acidalia and Cydonia regions. Near the south pole of Mars, a number of flat-topped hills are thought to be a group of tuyas that erupted under an ice sheet nearly a mile thick. Another type of feature that has been identified on Mars is a moberg hill, which forms when a lava flow intrudes into ice.

Despite evidence that surface ice was more extensive in the past, today only the poles of Mars have clearly visible ice deposits. This view was challenged recently when a group of British scientists analyzed new images of a region near the equator of Mars called Elysium Planitia. Previously, other scientists had mapped an area of terrain resembling broken, flat-topped ice floes, and interpreted it as a lava flow, reminiscent of platy lava flows in Iceland. However, new images of the area from the camera on the Mars Express spacecraft are much more suggestive of pack ice, rather than blocks of lava. If the geologists are correct, this frozen sea is about 500 by 560 miles, and possibly up to 148 feet deep. Crater counts on the dust-covered surface indicate it is about 3.7 million years old. It is certainly a likely target for future surface exploration of Mars.

The heat produced by a volcano erupting under an ice sheet or glacier will cause some portion of the ice to melt. This meltwater can sit for some time trapped in the ice, until enough of it accumulates to create enough pressure on the ice for a breakout. The result is a very sudden, massive flood of water erupting out of a glacier. This has happened often in Iceland, where the event is called a *jökulhaup*, a glacial outburst flood. The term can also be used to describe a flood produced by a melting glacier whose terminal moraine, the ridge of boulders and sediment that accumulates at the end of the glacier, fails under pressure from meltwater, also producing a massive flood. In Iceland in 1996, volcanic eruptions under the Vatnajökull glacier fed water to the subglacial lake Grímsvötn, producing a massive glacial flood that destroyed a major road, bridges, and laid waste to a large area. Luckily, the subglacial lake was being carefully monitored, and the area had been evacuated. Several eruptions have occurred since then, without melting enough water to cause a jökulhaup. Catastrophic floods on Mars may have been similar to jökulhaups, but most of the floods on Mars were caused by water or melting ice trapped beneath the surface, not on top as an ice sheet. When Martian permafrost suddenly melted, the trapped water escaped.

Near the equator of Mars in the Elysium region, the unusual platy texture of the surface is reminiscent of pack ice on Earth, which moves with currents and the wind. This has led some scientists to suggest that the area is a frozen sea that may have been liquid about five million years ago, with the ice still preserved under a layer of ash and dust. Other groups have likened the surface to some platy lava flows on Earth.

Fire and Ice When hot lava hits ice, it produces steam and water, which then have to go somewhere. If the ice is below the lava flow, the steam and water can explode through the lava, forming a feature called a rootless cone. The small cones on Mars (inset), located near the Olympus Mons volcano, may be rootless cones. Volcanoes can also erupt under ice, such as at the Vatnajökull glacier in Iceland (large image).

Snow Sleuth Author Tom Jones emerges (inset) from the Sierra Nevada Aquatic Research Labratory's snow lab on Mammoth Mountain (also an active volcano). Monitoring how the snowpack changes over the long term due to global climate change is critical—more than half of California's water comes from the Sierra snowpack. Global warming would result in snow levels retreating to higher altitudes, and an earlier spring runoff.

FIELDWORK:
SNOW TO WATER

To train the astronauts for our STS-59 and STS-68 missions, we took them into the field to meet our scientists and learn about the experiments planned for the Space Radar Lab. Two of the experiment sites were relatively close together, although they couldn't be more different: the dry desert floor of Death Valley and the snowcapped peak of Mammoth Mountain, both in California. The night the shuttle crew flew in, we were hit by a heavy snowstorm. We picked the crew up from an airport about 45 minutes away from Mammoth Mountain, and then had to drive on treacherous roads back to the site. The next morning, several feet of new snow covered the ground. We made it up by ski lift to the snowpack measuring station used by one of our investigators. Buried beneath feet of fresh snow and accessible only by a vertical manhole, the snowbound laboratory gave us access to snow layers of varying age and water content. The radar is very sensitive to ice, and the team could use radar images to estimate the equivalent water volume of the snowpack—a critical measurement for those who depend upon the yearly snowmelt. While the shuttle imaged the site from space, measurements were made on the ground to verify the space-based data. In the long term, being able to make global measurements of the snowpack on Earth will provide critical information not only on water supplies, but also on the changing climate.   —Ellen Stofan

The Future of Ice: *Climate Change*

SEE ALSO: *Layers of History* 136

THE CLIMATE OF EARTH has fluctuated over the history of the planet. A decrease in Earth's temperature of 7-9°F and an increase in snowfall marks the start an ice age. The generally accepted theory for why this shift occurs points to variations in Earth's orbit, first proposed by the Serbian scientist Milutin Milankovitch. Over intervals of 100,000 years, the eccentricity (or degree of circularity) of Earth's orbit changes, as well as the tilt of its axis. These Milankovitch cycles, when Earth receives less energy from the sun, correlate with the occurrence of ice ages. However, more recently, Earth's climate has been changing due to man-made, not natural, processes.

Earth's atmosphere contains a mixture of greenhouse gases, including carbon dioxide, that allow the sun's light to pass through, but trap heat radiated from Earth's surface. Earth's temperatures would be as much as 60°F cooler if we did not have greenhouse gases in our atmosphere. Venus has a "runaway" greenhouse atmosphere: The planet would have surface temperatures much more similar to Earth's were it not for its dense carbon-dioxide-rich atmosphere. The early Earth had a more carbon dioxide rich atmosphere and was much warmer, but then the carbon dioxide was trapped into carbonate rocks and reefs. Human activity since the industrial revolution, and especially in the last century, has injected increasing amounts of carbon dioxide into the atmosphere through the burning of fossil fuels. Given the well-understood effects of carbon dioxide as a greenhouse gas, are those rising levels actually enough to change Earth's climate?

The answer has come back from the scientific community as a resounding yes—humans are causing the climate to warm. This will lead to inevitable sea level rise as the warmer temperatures cause glaciers and ice sheets to melt. Current estimates suggest a global temperature rise between 1.98 to 11.52°F and a sea level rise from 7.1 to 23.2 inches during this century. While this might not seem like much, if we do nothing, it will be enough to flood highly populated low-lying areas on all the continents, change patterns of agriculture and disease, and determine whether certain animal and plant species survive.

In recent years, the effects of climate change seem to be accelerating. Ice has been disappearing from the Arctic and from Greenland at rates higher than predicted—in fact, the Arctic may be free of ice in the summer by the end of the century. Permafrost also has been melting at rapid rates in the Arctic regions, which in turn releases large amounts of carbon dioxide into the atmosphere due to its high concentration of organic material. While global warming cannot be stopped at this point, the worst effects may be mitigated by reducing carbon dioxide emissions. Technologies to sequester carbon are also being studied, and may help reduce the effects of climate change.

The chairman of the International Panel on Climate Change (IPCC) warned in his Nobel Prize acceptance speech that at least 30 percent of the species studied to date are at increased risk of extinctions as temperatures rise, and that number at risk could go as high as 70 percent if the average temperature rise is greater than 6°F (relative to 1980-1999). Not all regions will be affected in the same way; the Arctic, Africa, small oceanic islands, and the highly populated river delta regions of Africa and Asia are particularly vulnerable. For example, in Africa, the IPCC estimates that as many as 250 million people are likely to be subjected to water shortages due to drought caused by climate change.

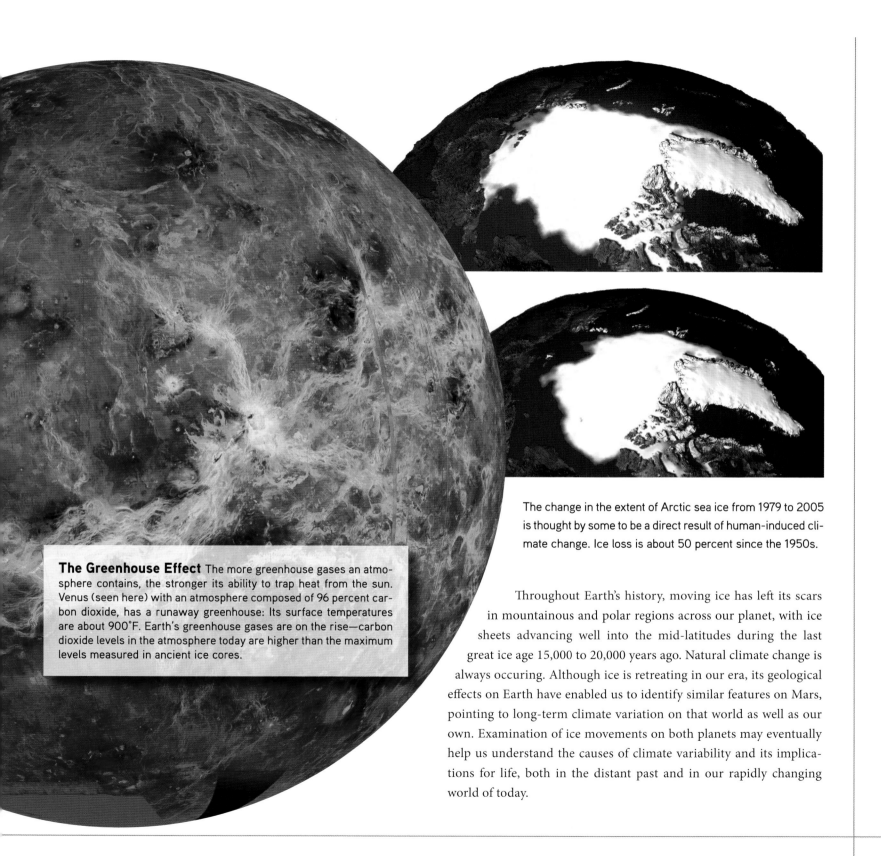

The Greenhouse Effect The more greenhouse gases an atmosphere contains, the stronger its ability to trap heat from the sun. Venus (seen here) with an atmosphere composed of 96 percent carbon dioxide, has a runaway greenhouse: Its surface temperatures are about 900°F. Earth's greenhouse gases are on the rise—carbon dioxide levels in the atmosphere today are higher than the maximum levels measured in ancient ice cores.

The change in the extent of Arctic sea ice from 1979 to 2005 is thought by some to be a direct result of human-induced climate change. Ice loss is about 50 percent since the 1950s.

Throughout Earth's history, moving ice has left its scars in mountainous and polar regions across our planet, with ice sheets advancing well into the mid-latitudes during the last great ice age 15,000 to 20,000 years ago. Natural climate change is always occuring. Although ice is retreating in our era, its geological effects on Earth have enabled us to identify similar features on Mars, pointing to long-term climate variation on that world as well as our own. Examination of ice movements on both planets may eventually help us understand the causes of climate variability and its implications for life, both in the distant past and in our rapidly changing world of today.

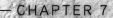

TOUCHED BY THE WIND

The wind sweeps clouds past the nearly one-mile-high volcanic peaks of Alexander Selkirk Island in the Pacific, then spirals inward to form eddies.

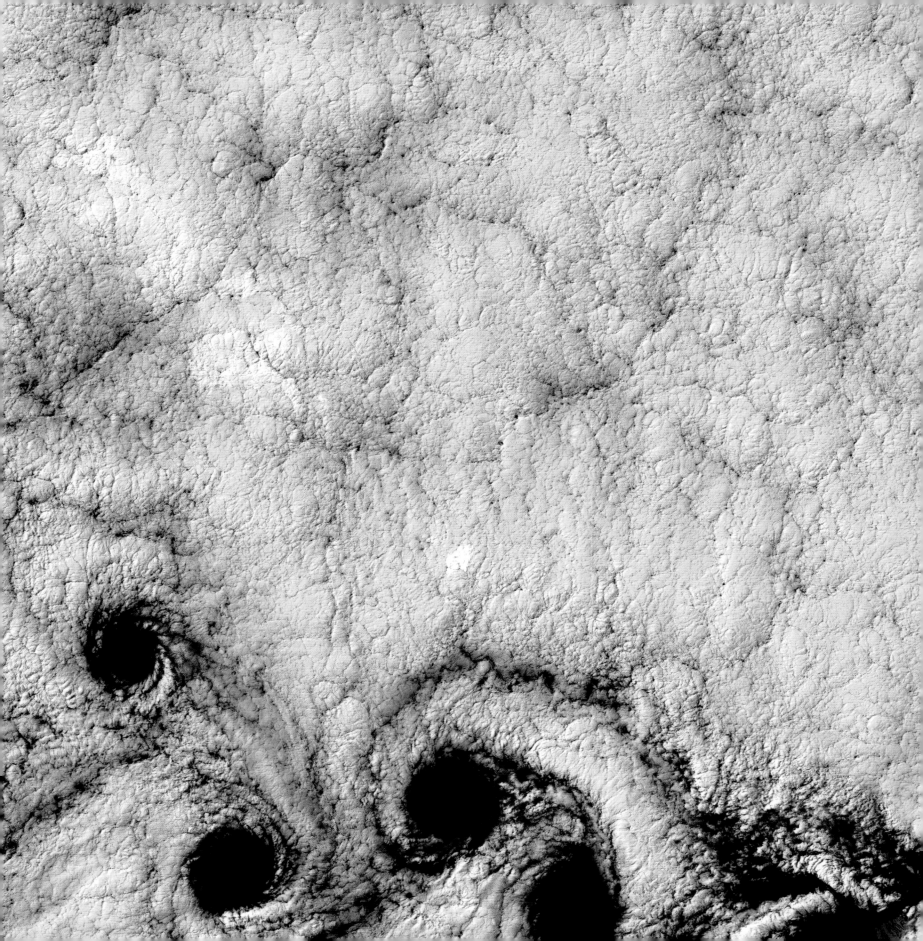

Indispensable Atmosphere

SEE ALSO: *Above It All* 159

THE WIND CANNOT raise continents or plunge oceanic crust into the abyss of the mantle, but it is a major force for reshaping our landscape. Earth's atmosphere is a nearly transparent film of air, so tenuous as to be overlooked at first. From orbit, we tend to look past and through our blanket of air to get at the *real* drama down on the surface. Ninety-nine percent of the atmosphere lies below 50 miles in altitude. Compared to the Earth's bulk, this gauzy blanket of air seems inconsequential. Yet it supports all life below, protecting us from meteorites, cosmic rays, damaging solar radiation, and the harsh temperature swings of deep space. For life on Earth, the atmosphere is indispensable.

This wispy envelope of gas around Earth is not just a passive insulator. Driven by solar heat, it has the mass and density to surge powerfully across the face of our world. Warmed in the tropics, heated air rises and spreads north and south, transporting solar energy toward the Poles. As it moves, displacing cooler air and deflecting under the planet's spin, the circulating atmosphere—the wind—can literally move the earth beneath.

The worlds of our solar system possess very different atmospheres. Venus is cloaked in a superheated carbon dioxide blanket so massive that the surface pressure is 90 times that of Earth's, equivalent to the water pressure found more than half a mile beneath our ocean. Intense solar heat and a runaway greenhouse effect heat the surface to some 900°F, hot enough to melt lead. Adding to the hostile climate are Venus's scorching upper-level winds, which circle the planet completely in just four days.

Mars, half again as far from the sun as Earth, has just a third of Earth's gravity, and is able to retain only a wisp of an atmosphere. Mars's average surface pressure is less than one percent of Earth's; its air is composed mainly of of carbon dioxide (95 percent) and nitrogen, with traces of oxygen and water. This thin envelope can barely protect against meteoroids and cosmic radiation, but it gets enough solar heating to raise strong winds and planet-wide dust storms.

Unlike the nearly bottomless atmospheres of Jupiter and Saturn, the latter's giant moon, Titan, possesses a thick atmosphere shrouding an enigmatic surface. Titan's mostly nitrogen atmosphere also contains a few percent methane and traces of more than a dozen organic compounds, such as ethane. The nitrogen, as on Earth, is in the form of N_2, and there's enough of it on Titan to create a surface pressure 1.5 times that of Earth's. The destruction of methane gas by weak sunlight creates a natural "smog" that shrouds the surface and produces a fine rain of organic particles onto the frigid landscape (-290°F). Surprisingly, Titan's sluggish winds have still managed to reshape its weird, ice-and-hydrocarbon surface.

On Earth, solar heating causes hot air to rise at the Equator and move poleward; cold polar air sinks and flows toward the tropics. Earth's rotation produces a Coriolis force that skews the moving air masses east and west.

Up in the Air Earth's atmosphere softens this view of the moon (large image), seen from the space shuttle. Unable to retain any true atmosphere of its own, the naked moon exhibits a crisp, sharp horizon against black space. Venus, which is the size of Earth but 30 percent closer to the sun, became hot enough to boil whatever oceans it once had. The massive carbon dioxide atmosphere with sulfur dioxide clouds (inset) trapped still more heat in a runaway greenhouse effect. The resulting surface heat baked the sensitive electronics of the U.S. and Russian robot landers into meltdown just a few hours after arrival.

An Invisible Force

SEE ALSO: *What's in the Wind* 166

THE INVISIBLE ATMOSPHERE is a surprisingly powerful force. Storms can wreak temporary havoc, but the force of steady winds, patiently shifting grains of sand and dust over thousands of years, can control and shape an entire region's landscape. On worlds with atmospheres, such as Venus, Earth, and Mars, and even Saturn's icy moon Titan, the wind's handiwork is everywhere in evidence.

Earth's atmosphere acts effectively to distribute solar heat from the tropics to the Poles. Composed primarily of nitrogen (78 percent), oxygen (21 percent), and one percent trace gases like water vapor and carbon dioxide, the atmosphere induces a mild greenhouse effect hospitable to both plant and animal life. Just one-millionth the mass of the planet itself, the atmosphere is still substantial enough, when the sun gets it moving, to lift soils and dust from the surface.

Storm winds, which have been recorded at speeds up to 231 miles an hour (clocked on New Hampshire's Mount Washington in 1934) are powerful enough to destroy forests and level man-made structures. Less dramatically, prevailing winds raise dust storms that carry sand and soil particles over long distances. These abrasive winds, although slower and subtler in action than water erosion, are nevertheless able to change the land around us.

And so it is on Mars. Martian winds are much weaker than terrestrial gales. Given the low surface pressure—less than one percent of Earth's—Martian winds must move about ten times faster to lift soils and dust. The energy comes from solar heating during the nearly six-month-long Martian summer: Rising hot air pulls in winds along the surface and puts the very fine, reddish dust in motion. Mars's lower gravity, only 38 percent of Earth's, also helps the dust get airborne and stay aloft long enough to migrate a long way.

When Viking landers touched down on Mars in 1976, the pair recorded average wind speeds of about 11 miles an hour, hardly enough to nudge a Martian sand grain. Yet lander cameras showed signs the wind had been at work: Drifts of fine-grained dust, like miniature sand dunes, and scoured faces of local rocks littered the landing sites.

Martian Atmosphere A Viking orbiter captured this view of the Martian atmosphere in the late 1970s. Just after the winter solstice, frost coats the surface of the southern hemisphere's Argyre basin, an ancient impact scar more than 1,100 miles across. Mars's atmosphere is 95 percent carbon dioxide and has a surface pressure less than one-hundredth that of Earth. Temperatures reach a high of about 70°F at noon at the equator in the summer, or a low of about minus 225°F at the poles. During the winter, as much as a third of the carbon dioxide condenses out of the atmosphere as "dry ice" frost. At the poles, the dry ice layer may be as much as three to six feet deep. Cold nights also produce a thin coating of water frost, seen by lander cameras, but it evaporates during daylight hours.

Titan's atmosphere is given its brown-orange tint by a haze of organic "smog" particles formed by the decomposition of methane by sunlight. Fallout coats the moon's surface and methane "rain" fills lakes near the poles. Titan's slow rotation (16 Earth days to complete a single Titan "day") induces just one sluggish Hadley cell, an air circulation pattern that circulates warm air from near the south pole to the north, while cold north polar air returns south along the surface. The Huygens probe, descending under its parachute in 2005, found wind speeds on Titan topping out at about 270 miles an hour around 70 miles up, but slowing dramatically near the surface. By the time the probe landed, the measured Titan breeze was dawdling along at just under one mile an hour.

But how could these weak winds be effective? Over hundreds of millions of years, repeated Martian storm winds stripped soil from some areas and piled it into vast fields of sand dunes. The gentle but steady action of Martian winds has even carved bedrock into fluted landforms aligned with the prevailing breezes. Orange-red dust, carried everywhere by the wind, is what makes the red planet red.

Wind Sculpts the Landscape

WHEREVER WIND FINDS LOOSE deposits of sand, soil, or dust (created by water-driven erosion or chemical weathering), moving air can alter the landscape. Where vegetation is absent or lack of moisture reduces the cohesion of soil particles, the wind can scoop up fine rock grains or soils in a process called deflation. This removal of local material can erase the original landscape. During the 1930s Dust Bowl, for example, drought and wind combined in places to remove more than three feet of the Midwest's topsoil in just a few years.

Deflation leaves behind shallow, saucer- or trough-shaped basins, usually just a yard or so deep and a few hundred yards across. These deflation hollows are very common in the semiarid Great Plains of North America. Astronauts on NASA training flights over Odessa Meteor Crater in west Texas observe so many sandy "blow outs" surrounding

Wind-borne sediments can sandblast away soft bedrock, leaving behind more resistant formations called yardangs. These yardangs in China's Yumenguan region, as much as 60 feet high, are some of the largest in Asia. Yardangs sandblasted into bedrock by steady Martian winds (right) were imaged near Olympus Mons in 2004. The flat region in the foreground is about ten miles across.

the impact depression that it can be hard to pick out the actual crater. Wind-driven soil has also mostly filled in the crater, once nearly a hundred feet deep.

Wind deflation typically occurs in deserts, on coastal beaches, lakeshores, and the loose floodplain deposits of large glacier-fed streams. During geology training in the valley of Arizona's Little Colorado River, astronauts have seen strong desert winds lift tons of fine sand into the air, and deposit it miles away on surrounding terraces in ranks of active dunes.

When the wind removes fine soil from a mixed surface of rocks and soil, heavier pebbles and stones are left behind. As the surface erodes, more and more pebbles remain at the surface, eventually producing a compact, stony surface that shields underlying dust from the wind. This new surface of concentrated pebbles is called desert pavement, and is common in Earth's deserts and even on the surface of Mars.

The wind can propel particles along in any of three ways: by rolling them along the surface, by lifting them into the air for short, parabolic hops, or, if the grains are fine enough, by carrying them aloft in suspension. Particles moving under saltation (a bouncing motion induced by wind) fall and strike the surface, dislodging other particles into the moving air. Water moves pebbles along stream bottoms the same way. About 95 percent of sand movement on Earth is by saltation. Once these particles get into the air, the moving soil, sand, and dust becomes an effective weapon of erosion.

This natural sandblasting can not only dislodge other soil particles but even attack solid rock. Deflation hollows can grow into wider depressions, separated by more resistant soil or rock surfaces. Erosive grit can eventually shape the resistant areas into a long, streamlined ridge, called a yardang (Turkish for "steep bank"). Yardangs typically have a sharp spine and resemble an overturned ship's hull; they can reach a hundred yards high and stretch along the prevailing wind direction for several miles. Yardangs are common in deserts around the globe, especially in the Gobi, and create striking landscapes on the dusty plains of Mars.

THE DUST BOWL

The 1931-39 Dust Bowl devastated the agricultural heart of the Great Plains. In 2004 NASA scientists exploring the Dust Bowl's causes found that weather changes in the Atlantic and Pacific tropics weakened the jet stream, diverting Gulf of Mexico moisture from the southern Great Plains. The resulting drought, coupled with the expansion of wheat farming that had plowed under the protective prairie grasses, stripped the land of vegetation and created the Dust Bowl. In the past, Midwest topsoils had been preserved during drought cycles by the hardy native grasses, but after the Civil War settlers plowed under large tracts of prairie for crop production. When drought caused the wheat harvest to fail, nothing could hold the soil against the relentless winds of the plains. Gusting winds across Kansas, Oklahoma, Texas, and Colorado gave rise to vast dust storms—"black blizzards"—that darkened the skies for days. In 1935 alone, 850 million tons of topsoil were blown away. Lost soil—an average of five inches over ten million acres—drifted around home-steads and across roads. Better soil conservation practices, tree planting, and a return of the rains in the fall of 1939 ended the Dust Bowl era. Although Midwest agriculture will surely suffer drought again, better farming practices, varied crops, and wooded buffers that reduce the impact of wind can help prevent the return of a dust bowl.

Desert Pavement The ruddy floor of Gusev crater on Mars was revealed by the Spirit rover's panoramic camera shortly after touchdown in January 2004. This desert pavement is probably created as wind transports fine soils and dust away, leaving behind a deposit of pebbles and stones. The wind-borne dust polished smooth the surfaces of these Martian rocks. Most of Africa's Sahara (Algeria, inset) is not covered with dunes, but by the same familiar pavement.

Dust Storms & Loess Deposits

ASTRONAUTS ON THE SPACE SHUTTLE frequently photograph dust swept from the Sahara out over the Atlantic Ocean; it often reaches the Caribbean. Research suggests that frequent dust storms in North Africa are linked to a decrease in the number and intensity of hurricanes that are formed in the eastern Atlantic Ocean.

Some Saharan dust, about 40 million tons each year, makes it all the way to the Amazon Basin. Windblown Saharan dust is an important source of mineral fertilizer for the Amazon ecosystem. Geologists have found that half of the annual dust supplied to the Amazon Basin emanates from a single source, the Bodélé depression located northeast of Lake Chad. Located in a narrow path between two mountain chains that direct and accelerate the surface winds over the depression, the Bodélé pumps an average of more than 0.7 million tons of dust into the air each day.

Satellite images of terrestrial and Martian dust storms show remarkable similarities in structure. But Martian dust storms dwarf even those legendary sandstorms of the Sahara and Gobi deserts. The typical dust grain size on Mars is just one micron across, as wispy as the particles in cigarette smoke, so even weak winds can turn the sky orange. Dust storms on Mars typically cover anywhere from 60 to 600 miles, but occasionally these local storms grow into mammoth, planet-wide shrouds of orange-yellow haze.

A 2001 dust storm engulfs refugees near the Afghan border at Chaman, Pakistan. One source of the dust is dry salt flats that were once the site of a lush oasis of lakes and wetlands near the border between Afghanistan and Iran.

When Mariner 9 arrived at Mars in 1971, nearly the entire surface was obscured by an opaque pall of dust. In 2007 dust storms coated the solar panels of the Mars rovers with enough dust to cut power to critically low levels. Fortunately, dust devils, regularly spotted by the rovers' cameras, have repeatedly swept grit off the solar cells, restoring adequate electricity production.

As wind speed drops, the suspended material falls to the ground, forming deposits of sand, soil, and dust. Steady winds can build up deposits called loess (German for "loose"), a thick blanket of fine dust, clay, and sand uniformly covering a region. These fine particles are very cohesive, and the grains in thick loess deposits seem to be almost cemented together. Loess is important on Earth because it forms rich farmlands in many areas, including the Pacific Northwest (the Palouse region of Washington State), the upper Midwest, and the Loess Plateau of central China. The source of loess is usually a nearby desert or the alluvial outwash plains of glacial streams. The ice ages left vast quantities of rock, ground to fine powder by the action of glaciers, exposed to scouring winds. Dust lifted from those sources created the large loess deposits in the interiors of North America, eastern Europe, and Asia. In eastern Nebraska, for example, the Missouri River carves steep-sided bluffs in loess deposits more than 60 feet thick. Similar deposits are found along the Rhine and Danube Rivers in central Europe.

Swirling Particles Springtime on Mars warms the swirled, carbon dioxide ice deposits of the north polar cap (large image). Vaporizing carbon dioxide raises the atmospheric pressure, enabling winds to lift dust more easily. Clouds of orange-tinted dust may stretch halfway to the Martian equator. On Earth, dust blown from the Sahara (inset), travels more than a thousand miles into the eastern Atlantic Ocean. Lofted as high as 15,000 feet, fine African dust can blanket a region as large as the continental United States.

ABOVE IT ALL

I n 1996, on the last day of the longest mission by a space shuttle, five astronauts aboard *Columbia* swept silently around Earth. Racing toward the horizon at Mach 25, we saw our planet's atmosphere as a thin blue layer of gauze, separating Earth from the utter black of space. Only along the Earth's limb—the horizon—can an astronaut gauge the thickness of that layer. From 220 miles up, the atmosphere is a thin blue haze, fading to black just a finger-width above the planet's curved limb. At night, the stars above, unchanging and cold, twinkle and shimmer as we watch them set through the atmosphere's turbulent and dusty layers. Looking closely, we could see this usually invisible air in action. Tropical islands, buffeted by trade winds, trailed delicate eddies of popcorn-shaped clouds. Distant storms etched the sea's deep indigo with lines of swells, as closely packed as the whorls of a fingerprint. On the southeastern U.S. coast, barrier islands of migrating dunes were pierced by turquoise channels, cut by recent hurricanes.

In the heart of the Algerian Sahara (left), vast seas of orange dunes, built by centuries of unceasing winds, stood in endless ranks. Other North African gusts lifted tendrils of tan and yellow dust far out into the Atlantic. In China's Gobi, powerful storm fronts churned sand into the air and threw it across the region, veiling thousands of square miles in a dismal yellow haze. In the Australian outback, blowing sand had carved long ridges into the fluted, rust-red bedrock. As we headed for home, strapped in for reentry, we five marveled at the tightly wound storm bands of Cyclone Daniella, whirling across the Indian Ocean to punish Madagascar with powerful squalls. There is no wind in space, but its gentle touch on our faces, and even its occasional furious attack, are some of the simplest pleasures of a return to Earth. —*Tom Jones*

Dust Storm on Mars

SEE ALSO: *Dust Bowl* 153

ARLIER LANDERS AND THE SPIRIT and Opportunity rovers found that the small-scale geology on Mars is dominated by a fine reddish dust. The average particle diameter is just one micron— 1/25,400 of an inch. Dust drifts on the downwind sides of rocks. It billows in small dunes, a yard or two across. Entrained in the winds, it sandblasts the rubble tossed out by meteorite impacts. And its red-orange tint coats everything, even the solar panels of machines from Earth.

Just as on Earth, the Martian atmosphere carries heat away from surfaces warmed by the sun. The heated, carbon dioxide "air" rises and moves poleward; cooler air descends and migrates toward the Martian equator to replace it. Because of Mars's rapid, 24.6-hour rotation, the Coriolis force, just as on Earth, causes an east-west as well as north-south circulation. Winds are also influenced by some very high Martian terrain: volcanoes that rise up to 15 miles above the surrounding plains, compared to Mount Everest's 6 miles above sea level. Martian winds

In June 2001, spring in Mars's southern hemisphere generated frequent dust storms as cold air from the south polar cap moved northward toward the warmer equator. Barely two months later, dust had shrouded nearly the entire planet; only the south polar cap remained clear.

June 10, 2001

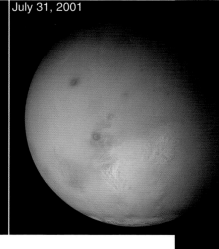

July 31, 2001

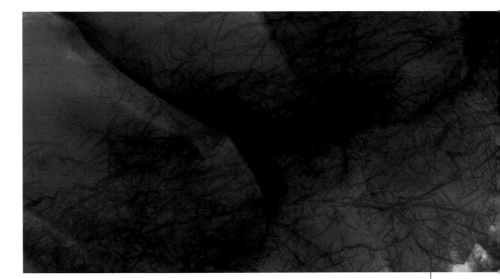

Mars Dust The solar panels of NASA's Mars exploration rover Spirit are so dusty that the rover almost blends into the red planet's landscape in this image mosaic taken by the rover's panoramic camera during October 2007. Dust on the solar panels reduces the rover's electrical output. Since Spirit landed on Mars in January 2004, passing dust devils have periodically blown the dust from the solar panels, restoring most power production. But the dust always returns: After enduring a major dust storm in mid-2007, Spirit's panels in 2008 were producing just 240 watt-hours of electricity, compared to the 900 watt-hours design level. Future Mars explorers, robot and human, will have to use nuclear power or develop practical ways of removing dust from solar arrays and vital equipment.

Dust devils are spinning columns of rising hot air—mini-tornadoes—which disturb fine soil as they spiral across the landscape. The Mars Global Surveyor caught these wild, looping tracks of Martian dust devils, here cutting through a dune field. The passage of each dust devil removes bright dust, revealing darker terrain underneath.

flow up and over these topographic highs and sink into the planet's low-lying plains and impact basins.

Robot explorers on Mars have measured a regular winter pattern of cold fronts passing overhead about every three days. As on our own planet, these are driven by the temperature differences between poles and equator, producing fast-moving weather fronts and gusty winds.

All of these winds can mobilize dust, but the big movers are huge summertime dust storms. The southern hemisphere, which is tilted toward the sun when Mars's elliptical orbit carries it closest to the sun, gets the hottest summers. That focused heating drives vigorous upper-level circulation toward the poles, inviting cooler air to rush in at the surface. Even in the thin Martian atmosphere, these winds can loft the fine surface dust high into the air, where it can stay for weeks at a time. The airborne dust absorbs more sunlight and warms the atmosphere's middle layers much more than surface heating can, intensifying the Hadley circulation toward the poles. These summer dust storms may arise at several places on the surface, then coalesce into a gigantic, planet-wide storm that chokes the skies for weeks. Wind-borne dust can thus travel across the entire planet. Some of it winds up at the poles; these layered dust deposits, locked in the polar caps, may preserve a record of dust storm activity over millions of years.

SEE ALSO: *Desert on the Move* 167

Dunes

THE MOST PROMINENT wind-borne deposits on this planet are the vast fields of sand that mark the heart of Earth's deserts. A sand dune forms where the wind velocity close to the ground drops enough (perhaps in encountering an obstacle) for entrained sand grains to fall to Earth. The growing pile of sand increases the wind disturbance and causes more sand to fall out. The prevailing wind piles sand on a gentle slope on the dune's windward face; as grains bounce over the top, they avalanche down the steeper rear, or slip, face.

Typical sand dunes can reach heights of 100 to 300 feet, but in the Alashan Plain of western China, some massive dunes are more than a thousand feet high. Dunes fall into five basic types: barchan, crescent-shaped with horns stretching downwind; transverse, with a sinuous ridge crest perpendicular to the wind; linear, with a ridge crest aligned between two prevailing wind directions; star, isolated sand hills with central peaks, built by multidirectional winds; and parabolic, U- or V-shaped with open arms facing upwind. Examples are found on Earth, Mars, Venus, and even on Saturn's moon Titan, wherever wind can deposit fine materials and drift them into hills. On Earth, so-called inactive dunes lie beneath present-day vegetation: the Sand Hills of Nebraska blanket about a quarter of the state and formed from sand eroded from the Rockies. If the local climate becomes drier, the grasses may die off, freeing these dunes to grow and migrate again. Where large amounts of sand accumulate, as in deserts, the dune fields can form extensive ergs, or sand seas. The heart of the Sahara and Australia's outback harbor such fields, with many examples also found on Mars.

Migrating dunes in coastal regions and along desert margins routinely destroy structures, bury roads, and cover productive farmland. Although some dunes can be stabilized by planting vegetation, halting their long-term migration is nearly impossible so long as winds and a source of sand persist.

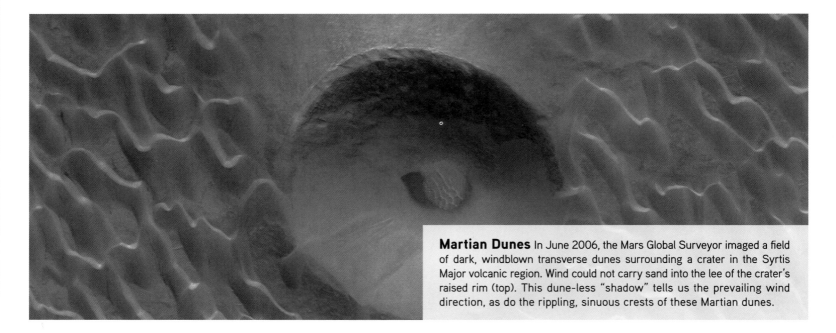

Martian Dunes In June 2006, the Mars Global Surveyor imaged a field of dark, windblown transverse dunes surrounding a crater in the Syrtis Major volcanic region. Wind could not carry sand into the lee of the crater's raised rim (top). This dune-less "shadow" tells us the prevailing wind direction, as do the rippling, sinuous crests of these Martian dunes.

TERRESTRIAL DUNE

Winds sculpt a 150-foot-high dune crest (large image) in the Algerian Sahara's Erg Bourarhet. Farther south, the Tifernine Dunes form the southernmost part of the Grand Erg Oriental, one of the driest, most desolate parts of the Sahara. In this STS-2 shuttle image (inset), barchan dunes, at the top, face prevailing winds from the west, and star dunes occupy the field's apex, where no wind direction predominates.

Otherworldly Dunes

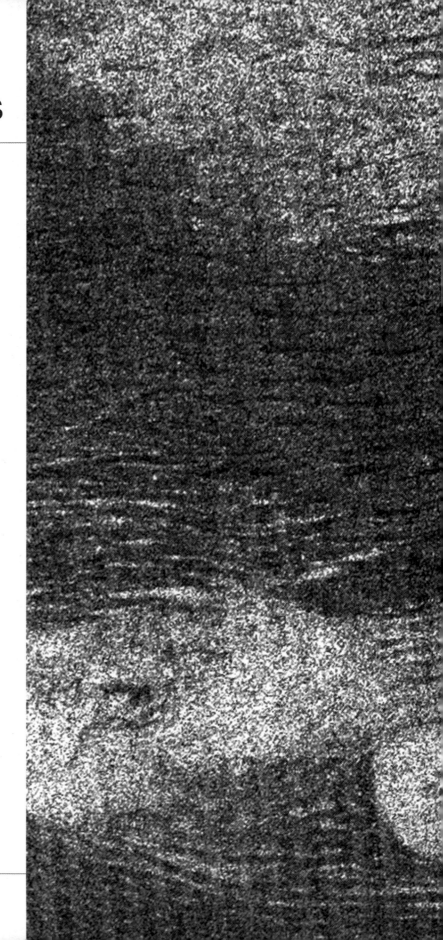

MARTIAN DUNES EXHIBIT all the classic terrestrial shapes, but barchan and transverse dunes are most common. Vast dune fields gird the northern polar regions, and the scarcity of impact craters there implies that these deposits are fairly recent. High-latitude craters near the south pole also trap wind-borne materials into dunes on their flat floors.

Some Martian dunes may reach heights of 20 feet or more. However, due to the thin atmosphere and low average wind speeds, dunes form and migrate very slowly. Under present conditions, sand takes some 50,000 years to drift into a classic dune shape. Winds must exceed 75 miles an hour to move even a three-foot-tall dune on Mars. And because peak winds are so infrequent, Martian dunes may take a thousand years to migrate just a few feet. Sand that forms dunes on Mars are similar in composition to sand on Earth—mostly small particles eroded from basaltic or sedimentary rock. The dunes are likely also coated with the very fine-grained martian dust that is blown around in the huge martian sandstorms that occur seasonally.

No one expected to find dunes in the frigid outer solar system, but Cassini's September 2006 flyby of Saturn's moon Titan captured this radar image (right) of long, dark ridges, thought to be longitudinal dunes on an area approximately 200 miles across. Geologists think that wind swirling around the bright, high-standing promontories produced the ridges, spaced about two miles apart. Instead of sand, these equatorial dunes may be composed of solid organic particles or ice coated with organic material. Solid organic particles rain down from the sky on Titan, formed by the interaction of methane and ethane in the atmosphere with solar radiation. Carl Sagan coined the term "tholins" for these particles, after the Greek word for "muddy." The particles are likely to be dark in color, but can range from sticky to fluffy in character.

Everywhere we've looked on worlds with atmospheres, we have found evidence of the wind at work. From Venus to Titan, dunes are the telltale signature of winds reshaping the planetary landscape.

Dunes and Streaks As on Titan (large image), dunes are also present on Venus. Wind streaks bend from southeast to west across this early 1990s Magellan radar image of the South Navka region of Venus (inset). At top right are fine lines of bright radar echoes, like iron filings bending around a magnet, probably Venusian sand dunes deposited by the region's steady, though sluggish winds. The inset image is about 60 miles across.

What's in the Wind

SEE ALSO: *An Invisible Force* 150

AS WITH WATER EROSION, wind-driven erosion and deposition are processes we see readily at work on Earth. On our home planet, hurricanes batter and flood the shoreline, drowning or displacing thousands; tornadoes can kill hundreds and wipe out entire towns; and dust storms lift millions of tons of grit into the air. Sand-blasting carves desert bedrock into fluted ridges, and the wind whips moving sand into marching ranks of dunes covering hundreds of square miles.

Wind is also at work nearly everywhere on Mars. Mars-orbiting satellites easily spot the tracks of dust devils on sandy plains. Wind has scoured dark soil from crater floors and blown it downwind, creating streaks stretching for dozens of miles. In its close-up inspection of the Victoria impact crater, the rover Opportunity has returned exquisite images of cross-bedded sandstone layers in the underlying bedrock, created from old dunes: The Martian winds have been at work there for many millions of years.

Our radar views of Venus and Titan have given us fewer details about their surface processes, but we plainly see dunes on Titan, and wind streaks and some bright radar echoes at Venus suggest strongly that dunes, while rare, do exist there. Every place in the solar system that has a solid surface and an atmosphere seems capable of producing sand dunes, evidence of the erosive power of wind.

Earth's atmosphere, with its greenhouse effect, typhoons and hurricanes, tornadoes, dust storms, and steady, rock-cutting winds, creates here all of the surface landforms we see on Venus, Mars, and Titan. Even our hurricanes are similar to, though much smaller than, the vast swirling storms in the atmospheres of the gas giants. Understanding our atmosphere's circulation, chemistry, and ability to modify the surface will help us not only with our investigation of the solar system but also will advance our efforts to preserve the precious envelope of air that keeps us alive.

This satellite image (above) from August 28, 2005, shows Hurricane Katrina's tight circulation and distinct eye. Katrina's interior wall is shown at right. Similar in character to giant cyclones on Jupiter and Saturn, the storm came ashore with winds estimated at 175 miles an hour.

DESERT ON THE MOVE

Desert terrain encroaches onto once viable farmland on the Ala Shan plateau, China (above). An estimated one-sixth of the world's population lives in desert climates. Such regions constitute some 15 to 20 percent of the world's landmass, and scientists are concerned that those areas are expanding. Our mental image of desertification is of sand dunes burying farms and creeping into village streets. But the term also includes destructive overuse of land in desert climates—overgrazing, soil exhaustion, and loss of topsoil through overtilling. The danger is that as the land degrades through abuse, people work it even more intensively to support farming or grazing activities, accelerating its exhaustion. The struggle by poor inhabitants to continue to extract a living from these threatened regions may subject adjacent, more marginal land to that same vicious cycle.

THE SEARCH FOR LIFE

From buffalo and egrets in Botswana to the insects and bacteria that we cannot see, Earth teems with life. Scientists have yet to find life on other worlds.

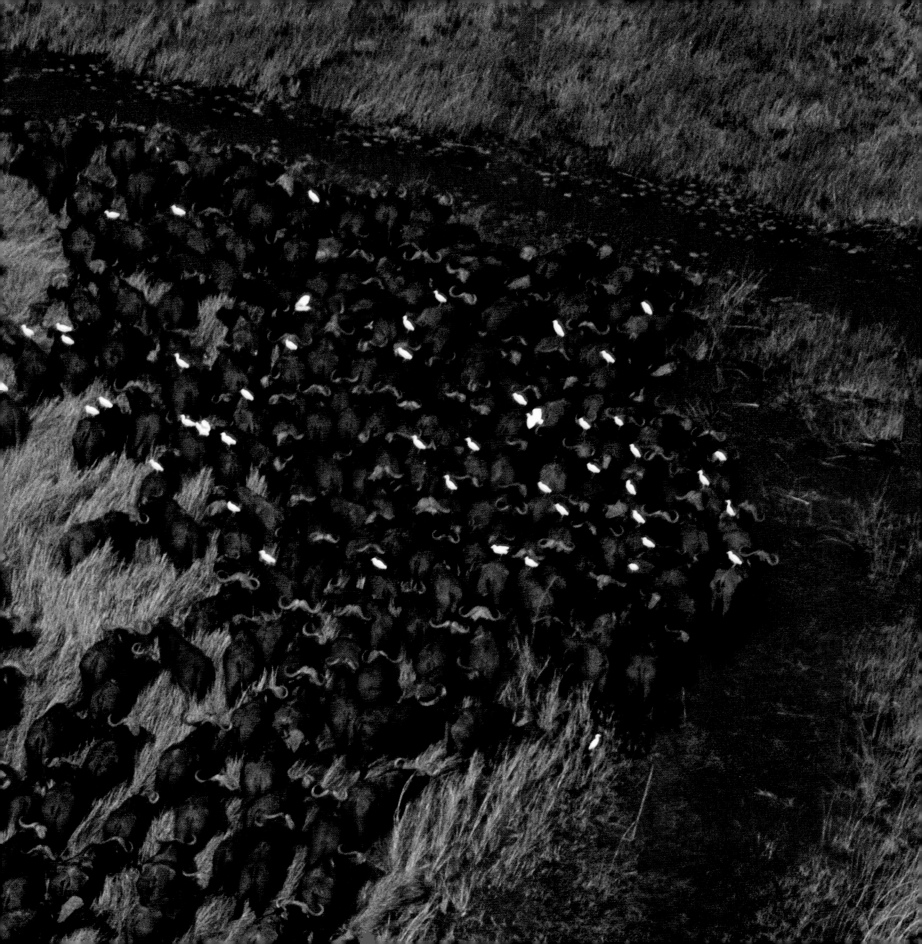

The Rise of Life

HUMANS HAVE long wondered whether we are alone in the solar system. From canals on Mars to aliens allegedly visiting Earth, our imaginations have outraced our actual knowledge. However, in 1996, one rock seemed to change all of that. A meteorite found in Antarctica (ALH84001—the first meteorite found in the Allan Hills region of Antarctica in 1984) was analyzed by a group of scientists in Texas. Their analysis produced multiple lines of evidence that life had existed on Mars; the most spectacular was the discovery of tiny wormlike features interpreted as fossil bacteria. In the years since the publication of their results, much of the scientific community has viewed the findings with skepticism. Other researchers claim that all of the features cited by the Texas group could be explained by chemical processes; life on Mars need not be invoked to produce the meteorite evidence. However, this work, combined with studies of life in extreme environments here on Earth, has led NASA to refocus its efforts on understanding the potential for extraterrestrial life.

Our understanding of how life evolves is understandably biased by the only example we have to study—our own Earth. There are three major components thought to be necessary for life: carbon, which along with hydrogen and oxygen make up the organic molecules that are the building blocks of life; liquid water; and a source of energy (the sun or heat from the interior of a planet). Life on Earth evolved in the presence of these three, and they are the three ingredients we look for when we search for life on other worlds. They probably need to be present in the environment for hundreds of millions of years to give life a chance to evolve.

The earliest evidence of life on Earth is fossilized traces of bacteria, found in rocks in western Australia that date back to about 3.5 billion years ago (the planet's oldest surviving rocks are 4.4 billion years old). A stable liquid ocean had formed as the early Earth cooled. The atmosphere, not as rich in oxygen, contained methane, ammonia, carbon dioxide, carbon monoxide, and nitrogen, along with other complex organic molecules delivered by comets. The denser atmosphere caused the surface pressure to be higher than it is today. Energy was introduced into this chemical environment in the form of lightning, ultraviolet light from the sun, and volcanic heat. We know from experimental work that chemical reactions of these ingredients under such conditions can produce more complex organic molecules needed for life, including amino acids. We have found some amino acids in meteorites, indicating that they can be formed in even harsher environments in the solar system.

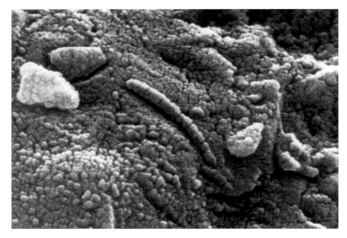

Are these microscopic specks fossils of ancient bacteria? Discovered on Martian meteorite ALH84001 with high-resolution scanning electron microscopy, some scientists believe these may be ancient signs of life. Others doubt these are actual fossils, as they are a hundred times smaller than any bacteria microfossils found on Earth.

3 BILLION YEARS AGO
Origins of microbial life—simple single-celled organisms

2.4 BILLION YEARS AGO
Oxygen becomes more abundant in Earth's atmosphere.

530 MILLION YEARS AGO
Cambrian "explosion" results in dramatic rise in complex animals in the fossil record.

400 MILLION YEARS AGO
Earliest insects appear in fossil record, along with several types of ferns and land plants.

AN ICY DELIVERY

Did life originate here on Earth or arrive instead from outer space? It's a question that seems straight out of science fiction, but in fact, it has been the subject of serious scientific inquiry. Scientists have detected organic molecules including amino acids in meteorites, and organic molecules have been observed in interstellar space. Organic molecules have even been found in several comets: Wild-2, visited by the Stardust spacecraft; Halley's comet, studied by the European Space Agency; and comet LINEAR. Scientists have also determined that organic molecules could survive the energetic process of an impact onto a planetary surface. Based on this evidence, we cannot rule out the hypothesis that some of the chemical building blocks of life arrived here from space.

350 MILLION YEARS AGO
Vertebrates established on land; first jawed fish emerge approximately 100 million years earlier.

225 MILLION YEARS AGO
Origin of dinosaurs following Permian extinction event 50 million years earlier

65 MILLION YEARS AGO
Extinction of dinosaurs occurs, probably precipitated by an asteroid impact.

4 MILLION YEARS AGO
Emergence of our earliest ancestors in Africa; modern humans appear approximately 100,000 years ago.

Kingdom of Life

WE HAVE SEVERAL THEORIES about how life formed from a collection of organic molecules. Common to all theories are complex sets of chemical reactions among these building blocks. One theory posits that RNA (ribonucleic acid—a critical component of all life on Earth) formed first, and then organized other molecules into groups and structures that eventually took on the characteristics of life and the ability to replicate and evolve. Liquid water was probably necessary, but not oxygen—Earth's early atmosphere had little of it. The sun was dimmer around four billion years ago, so Earth should have been cooler then. However, our atmosphere was richer in carbon dioxide, whose greenhouse characteristics warmed Earth and allowed a liquid water ocean to exist. As life evolved, it removed carbon dioxide from the atmosphere, to form shells and coral reef structures, resulting in an atmosphere that eventually became more oxygen-rich. Once life did gain a toehold, single-celled life forms (mostly bacteria) dominated until about a billion years ago, when multicellular organisms evolved. So even though life took hold relatively rapidly on the early Earth, it took billions of years for life to evolve beyond relatively simple forms. Land plants first appeared about 400 million years ago; dinosaurs dominated Earth between 230 and 65 million years ago. After the dinosaurs died off, flowering plants appeared and mammals, which first evolved about 215 million years ago, flourished and diversified. Finally, about four million years ago, early humans appear in the fossil record in Africa.

Why do species evolve and then become extinct? Living organisms change or evolve as they adapt to changing conditions. Specifically, species change as their DNA mutates, with natural selection promoting the spread of mutations that benefit a species. For example, some bacteria evolve to become resistant to antibiotics. In general, the process of a species evolving into an entirely new species takes place over long timescales. Species become extinct when their ecological niche changes, and they cannot adapt. These changes can occur through events like a catastrophic impact or natural or human-induced climate change. Species can also become extinct when they evolve into a new species. Abundant evidence of the extinctions and evolution of species are found in the fossil record on Earth and can also be studied in the laboratory. The story of life on Earth has been long and successful, but incredibly complex. Simple life-forms evolved relatively rapidly—bacteria still represent the most common life-form on Earth. More complex organisms such as reptiles and mammals appear

Organisms have adapted to life in nearly every environment on Earth. The largest living organism, a fungus called *Armillaria ostoyae* (honey mushroom), has been found in Oregon and Michigan, covering thousands of acres. It lives underground, occasionally appearing above ground as a small, gold-colored mushroom.

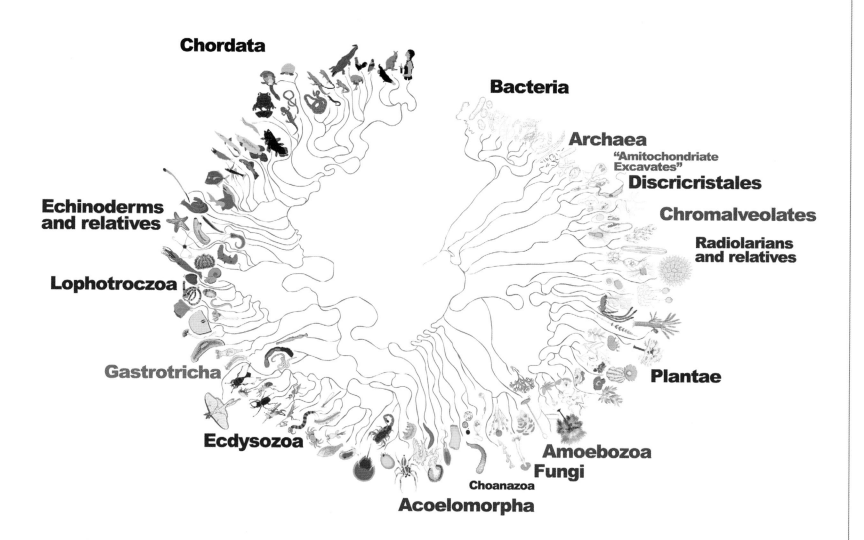

Chordata

Bacteria

Archaea

"Amitochondriate Excavates"

Discricristales

Chromalveolates

Radiolarians and relatives

Echinoderms and relatives

Lophotroczoa

Gastrotricha

Plantae

Ecdysozoa

Amoebozoa

Fungi

Choanazoa

Acoelomorpha

very late in the history of Earth, with millions of species coming and going as conditions on Earth changed. In the case of mammals, more than a hundred million years passed between when they first appeared in the fossil record and when they became more widespread (after the impact-induced fall of the dinosaurs at the end of the Cretaceous period 65 million years ago). Life not only requires favorable conditions to evolve at all but can only flourish and expand with stable ecological niches, enabling life to persevere over time.

The Tree of Life shows the evolutionary relationships between all life on Earth. Charles Darwin was the first to use the analogy of a branching tree to describe the connections between extinct and living species.

As we search for life on other worlds, we must answer two separate questions: First, did the conditions exist to allow life to evolve, and second, do these conditions last over the very long time periods that enable life to persist and diversify?

Life in the Extreme

SEE ALSO: *Hot Springs, Mars and Life* 179

THE CONCEPT OF LIFE ON OTHER WORLDS moved out of the pages of science fiction as we learned that life on Earth is much more tolerant of extreme conditions than we had ever guessed. Scientists had thought that life on Earth could withstand only a relatively limited set of conditions—an average temperature of 56°F, in relatively mild pH conditions (not too acidic, not too basic). We now know about an entire class of organisms called extremophiles; organisms that flourish under extreme conditions. Extremophiles belong to the domain archaea in the tree of life. The conditions in which they live include extremes of temperature (hot thermophiles and cold psychcrophiles), isolation from the surface of Earth (endoliths—organisms that live inside rocks), very acidic (acidophiles) or very basic (alkaliphiles) conditions, or without oxygen (anaerobes). There are even organisms that survive and thrive in very dry environments, very salty conditions, under very high pressure, or even immersed in nuclear waste. When scientists visited places like the very acidic Tinto River in Spain, the hot spring pools at Yellowstone National Park, undersea volcanic vents, and lakes below the ice in Antarctica, they invariably found life. On our Earth, life seems to continually redefine the concept of habitable environments.

Life on Earth is amazingly diverse. Fewer than two million species have been classified on Earth, but it is thought that many millions more have yet to be classified. We know of an 8,500-year-old fungus that extends over nearly four square miles (the largest known living thing on Earth), and bacteria that can exist at 239°F. Perhaps the most humbling realization for humans is that we are so outnumbered. Scientists estimated recently that there are 5×10^{30} bacteria on Earth, a number that makes the 6.6 billion (6.6×10^9) of us completely insignificant! The important lesson is that simple forms of life are the first to evolve and are likely to be the most abundant on any planet. As we search for life in the solar system, our best bet is to focus not on the alien creatures most people think of as extraterrestrials, but on microscopic evidence that life has found another home.

Methane Ice Worms On the floor of the Gulf of Mexico, these one- to two-inch-long worms (large image) live in piles of methane ice. The worms may eat bacteria that grow on the methane hydrates that form under the high pressure of the ocean floor where subsurface deposits of oil and methane gas seep up through cracks. The worms were discovered using a mini-research submarine deployed by a team of scientists studying the deep ocean floor for the National Oceanic and Atmospheric Administration. Other hostile environments on Earth, such as the acidic Tinto River in Spain (inset), are also host to extreme forms of life.

1 m m

SEE ALSO: *Venus: The Once and Future Earth?* 182

Where Do We Look?

OUR SEARCH FOR LIFE has understandably been biased by what we have learned here on Earth. Terrestrial experience has us looking for environments with liquid water, a source of energy, and the chemical building blocks of life. We have detected organic molecules on comets and in meteorites, and we know they have been delivered all over the solar system, so the last criterion is not a hard one to meet. In general, neither is the second; most bodies we have visited in the solar system exhibit evidence of past and/or present volcanism. The limiting condition in our solar system appears to be the presence of liquid water, which must be stable on or near the surface for millions of years. Water's relatively narrow range of liquid stability forces us to look either into the past (for Mars and Venus) or under the surface (Mars and Europa).

It is not only important where we look, but also how we look. The search for life in the solar system takes many forms, from the array of telescopes that both collects radio astronomy data and searches for signals of intelligent, extraterrestrial origin to orbiters and landers at Mars and the moons of Saturn. To date, the search for extraterrestrial life focuses on indirect rather than direct evidence. Spacecraft search for evidence of liquid water, now or in the past, by studying surface landforms, the chemical composition of surface rocks, and the composition of the atmosphere. Gravity data and rotational information are used to detect subsurface oceans. Environments in which life could evolve, such as subsurface oceans on Europa, Titan, and Enceladus, have been identified by the Galileo and Cassini spacecraft. Future landers at Mars and the icy outer planets will search for water and surface composition, and ultimately attempt to detect life directly.

The Allen array of telescopes, located at the Hat Creek Observatory in northern California, is run by the SETI Institute and the University of California, Berkeley. Currently consisting of 42 arrayed telescopes, it will have 350 telescopes when completed. It is used to search for radio signals from extraterrestrial sources, as well as astronomy research.

MARS

Mars has dry riverbeds on its surface that indicate a watery past. An array of spacecraft have been studying the history of water on Mars and have found some intriguing results.

EUROPA

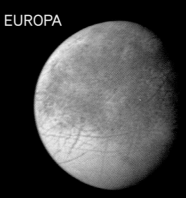

Europa, one of Jupiter's 60-plus moons, is similar in size to Earth's moon but composed mostly of water ice. Its complex surface overlies an ocean; how far down is unknown.

TITAN

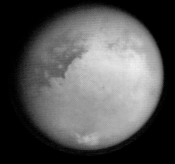

Haze shrouding Saturn's moon Titan is partially composed of organic particles that form in the atmosphere and rain onto the surface. Though cold, it may have the building blocks of life.

Life on Mars?

SEE ALSO: *The Rise of Life* 170

MARS HAS LONG BEEN REGARDED as the leading candidate for finding life beyond Earth. Its riverbeds and giant volcanoes indicate that two of the critical ingredients have been present: water and a source of energy. Geologic features indicate that water might have existed on the Martian surface for a long time. The more narrow valley networks in the southern hemisphere indicate that, at least early in Mars history, there were periods of rainfall. Snow must also have fallen, based on the evidence for Martian glaciers. Other evidence supports the idea that for long periods of Martian history, most of the water existed as ice in the upper layers of the crust. Large outflow channels suggest that Mars underwent sudden outbursts of massive amounts of water, perhaps caused by an asteroid impact or a large volcanic eruption that melted subsurface ice.

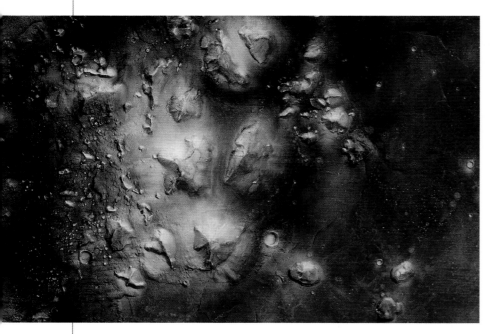

A small mesa with humanlike features gained fame as the "face on Mars" when imaged by NASA's Viking spacecraft. Better imaging shows that geologic processes, rather than Martians, carved this landscape.

THE SEARCH **FOR LIFE**

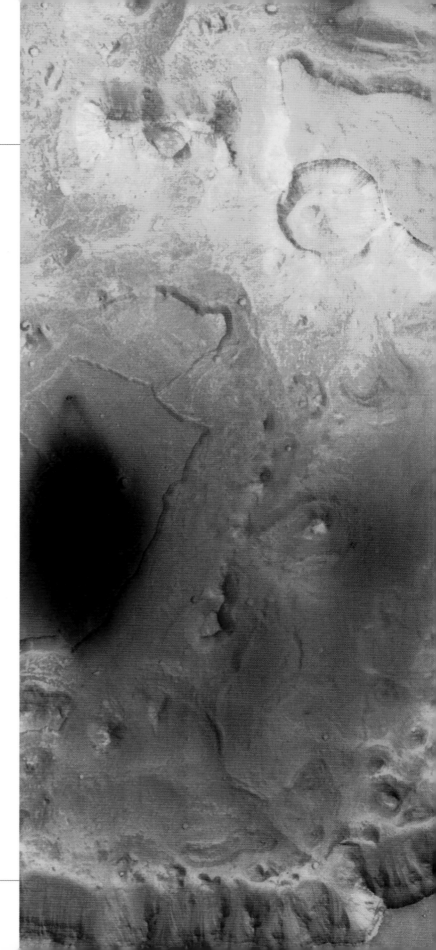

Hot Springs, Mars, and Life The Valles Marineris canyon shows evidence of having been eroded by flowing water (large image). To find evidence of past life on Mars, the best places to search are where water may have persisted for a long time on the surface, such as ancient hot springs. On Earth, hot springs form at places like Yellowstone, Wyoming (inset), when magma close to the surface heats underground water. In 1966, Dr. Thomas Brock discovered the first thermophile bacteria in Yellowstone. Now more than 50 species of heat-loving bacteria have been identified, and many think a hot springs type of environment is where life began on Earth. To find ancient hot springs on Mars, researchers look for geochemical and geomorphologic signatures such as the terraced mounds at Mammoth Hot Springs.

The U.S. and European Mars exploration programs pursue a "follow the water" strategy to enable us to better understand the history of water on the planet. Water is the key to understanding life on Mars; it plays a critical role in the geologic and climate history and provides an important resource for future human missions to Mars. Measurements from orbit suggest that two particular regions on Mars (Gusev crater and Meridiani Planum) were water-rich in their past. This led to more detailed studies of these regions by the Mars explorations rovers Spirit and Opportunity. In the Meridiani area, Opportunity found geochemical evidence that very acidic, water-rich conditions existed and found rocks possibly formed by sediments deposited in flowing water. The Spirit rover, in Gusev crater, also found minerals that must have formed in the presence of water, though water does not seem to have been as abundant in this region. While neither rover has found any evidence of life, Meridiani's environmental conditions may at some point have been able to support life.

Europa's Watery Secrets

SEE ALSO: *Volcanoes of Io* 89

I N CHAPTER 4, WE SAW HOW Jupiter's tidal pull results in a very volcanically active moon, Io. Io is composed mostly of silicates, so the tidal heating melts the interior, creating magma that erupts onto Io's surface. On Jupiter's icy moon Europa, the tidal pull of Jupiter also melts the interior, but Europa's interior is largely water ice. The Galileo spacecraft measured that moon's gravity field, indicating the presence of a liquid ocean somewhere beneath the surface—a water layer that could be more than 60 miles thick. In addition, images of Europa's surface showed evidence for cryovolcanism, further evidence of a subsurface ocean. The cryovolcanism and that subsurface ocean provide the water and a source of energy that would be essential to life. Because we know that organic molecules are abundant in the solar system, they are also likely to be present on Europa.

How far beneath Europa's surface is the ocean? At times, through volcanic eruptions, the liquid water breaks through to the surface. However, the surface of Europa is bathed in intense radiation from Jupiter so strong it would kill any living organism; we must search below. Estimates of the thickness of the outer crust range from about 6 to 15 miles, although some argue it could be much thinner in places. Crater shapes do not change much around the moon, suggesting that the thickness of the crust is fairly uniform across Europa. Other scientists have suggested that the unusual chaos terrain and some of the ridges on Europa means that the ocean must occasionally "leak" onto the surface. If we are to ever assess the potential for life on Europa, scientists must find out if indeed there are thin places on the crust where the ocean can be sampled. NASA has studied ideas for entering the ocean below searching for signs of life: A "cryobot" might melt through the crust of Europa and cruise through the deep, pitch-black waters.

The most likely type of life on Europa would be single-celled life-forms that require no sunlight to survive. Europa's deep-ocean organisms would have to use the available chemistry to find nutrients: The deep waters may support ongoing chemical reactions similar to those in Earth's hydrothermal environments, where mini-ecosystems thrive today.

The Conamara region of Europa shows blocks of icy crust that have broken up and moved around like pack ice, suggesting that Europa's subsurface ocean may be near the surface in this region.

HOT SMOKERS

Deep-sea or hydrothermal vents are produced by magma beneath the ocean surface coming into contact with underground water, causing episodic or even continuous eruptions of extremely hot, mineral-rich water. They are actually geysers on the ocean floor, supporting a wide array of life. There are two types of vents: black smokers and white smokers. Black smokers have much hotter erupting fluids, so hot they dissolve metals in the fluid, which then react with seawater to form black particles. As a vent erupts, a tall chimney of mineral deposits builds around the initial eruption vent, such as these photographed on the floor of the Pacific Ocean. These chimneys can grow to be several hundred feet tall. The first hydrothermal vent was discovered in 1977 in the Pacific Ocean. Now hydrothermal vents have been found in both the Atlantic and the Pacific.

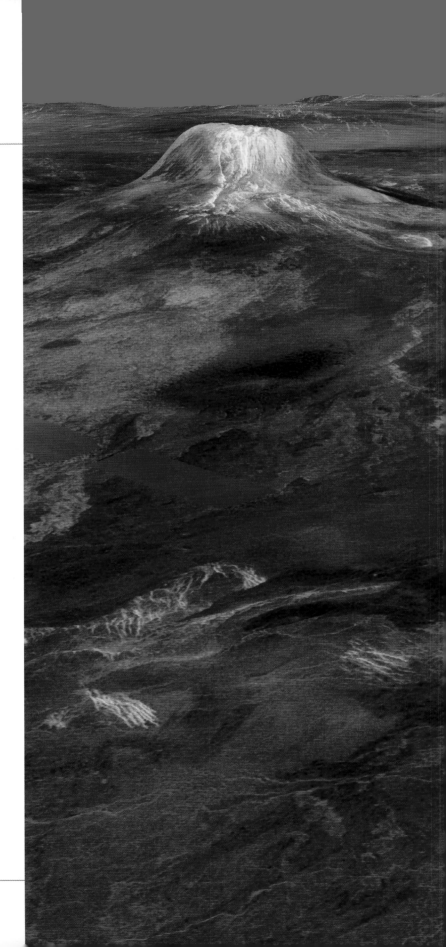

Venus: *The Once & Future Earth?*

LIFE ON VENUS? NOT POSSIBLE with those 900°F surface temperatures, high pressures, and dense carbon dioxide atmosphere. But was Venus always the inhospitable place it is today? Research into the history of the Venusian atmosphere has indicated that a liquid ocean may have existed on the surface of Venus early in its history—possibly for as much as a billion years. During that time span, simple, single-celled life-forms evolved on Earth. Could the same have happened on Venus? Better geochemical measurements of the surface rocks and the atmosphere of Venus would help to answer this question; such measurements would focus on chemical tracers in the atmosphere that help tell us more about the planet's history. The chemistry of surface rocks would also zero in on the history of water on the planet, because water's presence has very specific effects on rock composition. As with Mars, the best evidence for past life on Venus would be fossils; a rock that actually preserved signs of past biological life. However, on Venus this search would be like finding a needle in an impossibly large haystack: Building spacecraft that could conduct such a search is beyond our present skills. Worse, the planet's active tectonic and volcanic past has probably destroyed or covered up most rocks from the very earliest portion of the planet's history. Most of the surface of Venus that we can see using spacecraft radar images has formed in the last 750 million years, long after water likely disappeared from the planet. Where did the water on Venus go? As the planet became hotter early in its history, the oceans would have begun to evaporate. Water vapor is a greenhouse gas, so temperatures in the atmosphere would have increased, resulting in more evaporation and even higher temperatures. The process would have continued in a runaway greenhouse effect, until all of the surface water had evaporated. The hydrogen and oxygen in water molecules in the atmosphere would then break apart by reactions with other gases and due to the effects of sunlight, with the hydrogen escaping into space. Luckily, our greater distance from the sun results in lower temperatures here on Earth, limiting the amount of water vapor the atmosphere can hold.

A Dry World This view of two volcanoes (large image) on Venus was created by draping radar images over topography, and then coloring the images to mimic the filtering effects of the thick Venusian atmosphere. On Earth, even in dry regions we find evidence of past water, such as this dried-up surface from the badlands in North Dakota (inset). However, on Venus, any ancient lake beds are likely covered by lava flows or broken up into complex tectonic terrains. To find evidence of water in Venus's past, we must send a lander to the oldest rocks we can find, as they may preserve clues of a wetter past.

GORILLAS

On the border separating Rwanda, the Democratic Republic of the Congo, and Uganda lies the Virunga volcano chain, home of the endangered eastern gorilla subspecies, the mountain gorilla. These gorillas were studied by the late Dian Fossey (1932-1985), whose work was popularized in the movie *Gorillas in the Mist*. During the first Space Radar Lab (SRL) mission in April of 1994, Tom was on orbit, and I was helping organize science operations from the ground at the Johnson Space Center in Houston, Texas. Just before the flight of the shuttle *Endeavour*, we got a call from Dr. Scott Madry, who was working with the Dian Fossey Gorilla Fund, asking if it was possible for us to use the radar to image the area around the Virunga volcanoes where the gorillas live. They wanted to use the radar to study the vegetation patterns that would provide clues to the extent of the gorillas' habitat in this remote and rugged region. We were thrilled to be a part of helping an endangered species, and quickly figured out how we could gather the data they needed, both on our first mission and our second flight in the fall. This image we took is my favorite of the whole mission, not only for its value in understanding the endangered gorillas but also in the contrast between the dramatic volcanoes and the surrounding terraces of vegetation. The false color radar image shows a 36-by-35-by-107-mile area in the Virunga region, and was made by combining two bands of radar data from the first flight of the SRL mission. The black feature to the right is Lake Kivu, which forms part of the border between the Democratic Republic of the Congo and Rwanda. Mount Karisimbi (14,764 feet) is in the center of the image. —*Ellen Stofan*

Endangered Gorillas There are only about 700 mountain gorillas left, all in an extremely war-torn part of Africa. They have been hunted, captured for zoos, decimated by diseases and habitat destruction, and caught in the middle of human warfare. Wildlife groups, particularly the Dian Fossey Gorilla Fund, have worked to fight poaching and preserve gorilla habitat. The center was destroyed after the 1994 war in Rwanda, but its work continues from a new base near the Virunga Volcanoes Park.

Titan & Enceladus

TWO MOONS OF Saturn, Titan and Enceladus, offer additional possibilities as sites of life in our solar system. Both moons have the required ingredients—organic molecules, water, and a source of energy—that scientists Stanley Miller and Harold Urey studied in their famous experiment. On both Titan and Enceladus, the likely form of energy is cryovolcanism produced by heating in the interior of the moons. Both also have a major drawback—extremely cold surface temperatures, due to their distance from the sun. Saturn has an average solar distance of about 886 million miles, resulting in surface temperatures of minus 290°F at Titan and minus 320°F at Enceladus. While these temperatures are much colder than those at which we think life can exist, these moons might have subsurface oceans where temperatures would be warmer. We know that liquid water can exist in very thin films in rocks down to minus 20°F, and bacteria has been found living at temperatures as low as this on Earth, but our understanding of life's low-temperature limits could be biased by how life exists here on Earth.

On Titan, particularly intriguing places to search for the organic molecules that might precede life are the lake-rich north and south polar regions. It has been suggested that the dunes in the equatorial region of Titan could be formed of organic particles that have rained out of the atmosphere. Titan has similar chemistry to that which characterized early Earth, with the exception of a carbon dioxide–rich atmosphere. The combination of possible cryovolcanism on Titan with its organic-rich surface makes it a tempting target. Titan still may be too cold for even simple life-forms, but the organic molecules we study there may help us better understand the reactions that led to the formation of life on Earth. In addition, analysis of Titan's orbital characteristics has led the Cassini science team to propose that Titan has a subsurface liquid ocean, similar to that on Europa. In this warmer subsurface liquid environment, similar to Europa's, life may be possible. Further analysis will help us better understand these potential environments for life.

Cassini scientists were aware of large cracks on the surface of Enceladus from images taken by the previous Voyager mission to Saturn. However, they were surprised when Cassini's instruments detected anomalously warm temperatures over the cracks and captured an image of an erupting geyser. The Cassini spacecraft passed over the cloud coming from one of Enceladus's erupting geysers and detected water vapor, carbon dioxide, carbon monoxide, and organic materials. So, Enceladus has organic-rich chemistry and an internal heat source producing subsurface liquid water. Liquid from the interior erupting onto the surface means that subsurface layer should be accessible to future probes.

If life did exist on Titan or Enceladus, what kind of life would it be? Extremophiles here on Earth provide some clues. There are methanogens, which can live without oxygen; other bacteria that live off the radioactive decay of elements in rocks; and bacteria that live off the energy from chemical reactions between different rocks. Another possibility is that the subsurface oceans on Titan and Enceladus could have hydrothermal vents, at which life could thrive.

Stanley Miller (above) and Harold Urey produced amino acids—the building blocks of life—by combining water, methane, ammonia, and hydrogen with evaporation and condensation, and introduced sparks to reproduce lightning. After a week, amino acids formed.

Return to Titan Scientists would like to return to Titan to further investigate its rich organic environment (below, artist's concept of the 2005 Huygens lander of the surface). However, travel to Titan takes at least seven years! A new mission would likely consist of an orbiter to produce a global map of its surface composition and better knowledge of its internal structure, and a lander to investigate the organic compounds at the surface. The possibility of sending some sort of balloon outfitted with instruments that would float above the surface is also being studied. The lakes on Titan are a possible target for future landers, as are the organic-rich dune fields near Titan's equator.

The Search Continues

SEE ALSO: *In the Zone* 204

LIFE ON EARTH IS ABUNDANT and varied; the life-forms we might find on other bodies in the solar system may be quite simple, single-celled organisms. Earth lies in the so-called habitable zone, the perfect distance from the sun where liquid water is stable on the surface. Other bodies on the edges of the habitable zone, such as Mars and Venus, may have had liquid water stable on their surfaces long enough for life to have evolved. It may still exist in some isolated environments, especially on Mars. Farther out in the solar system, conditions favorable to life may exist in water oceans below the icy crusts of Europa, or possibly even on Enceladus or Titan. The search for life or evidence of past life in our solar system is ongoing. In the next chapter, we describe our search for other Earths: planets around other stars that may also have environments suitable for life.

The story of life on Earth is one of change—new species evolve and others become extinct. Yet Earth has environments like this coral reef in the Red Sea in which the sheer diversity of life is impressive—it is these environments we seek elsewhere in the solar system and the universe.

NEW WORLDS

The Hubble telescope reveals clouds of hydrogen gas, collapsing under intense starlight. This stupendous pillar may be an incubator for newborn stars and planets.

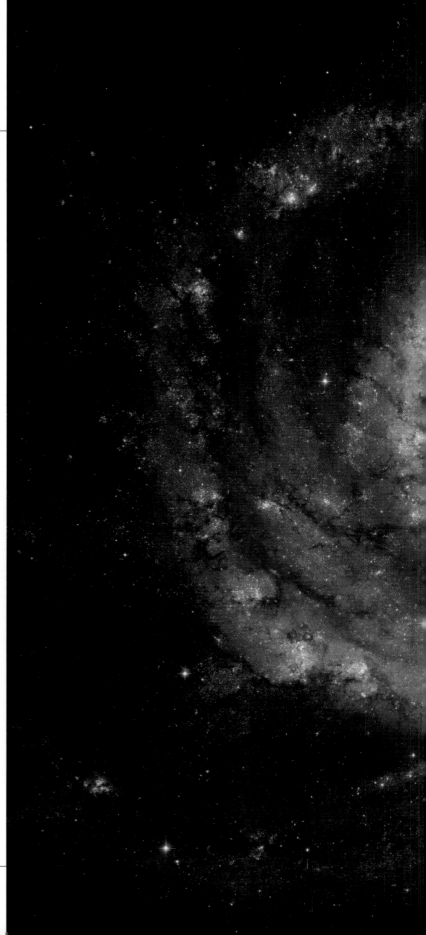

Shaking Up *the Neighborhood*

SEE ALSO: *The Planetary Zoo* 202

WHETHER IN SPACE or here on the ground, we humans are residents of our own little solar system. Our planetary neighbors range from hot to cold, small to big, rocky to icy, even from dynamic to dead. To our knowledge, only one of these worlds harbors life.

We have long suspected that some fraction of stars in the Milky Way also hosts families of planets, but for millennia the sun's family was all we could know and study. Our telescopes just weren't capable of detecting another world trillions of miles away. But in 1995 astronomers announced the discovery of a planet circling the star 51 Pegasi, 50 light-years from Earth. This new world, 150 times the mass of Earth, hugs its parent star so closely that it completes an orbit once every 4.2 days. Its tight orbit means that the planet A51 Pegasi-b is scorched by the searing heat of its sun.

But this uninhabitable world is nevertheless a planet, and it broke our sun's lock on its exclusive status as a parent star. If one world is out there, there must be others—in all probability, millions, if not billions.

Dark tendrils of dust, backlit by millions of stars, shroud the center of our galaxy, the Milky Way, seen here from Easter Island in the Pacific.

Galaxy M51 The famous Whirlpool Galaxy, trails two spiral arms alight with billions of stars. Thirty-one million light-
years away, the arms are busy triggering the birth of new
stars by compressing hydrogen gas. Hubble imaged the bright
pink star-forming regions in January 2005. Intense blue traces
young star clusters on the outer margins of each arm, some of
which are circled by newborn planets (artist's concept, inset).

Playing Hard to Find

SINCE COPERNICUS PROPERLY placed Earth in its elliptical path circling the sun, just one member of a family of planets, astronomers have sought evidence for other solar systems around other stars. But as the true scale of our stellar neighborhood became known, the hopes of early telescope-wielding astronomers faded. The nearest star, Proxima Centauri, is 4.22 light-years away, more than 24 trillion miles. At that distance, our instruments could neither distinguish the tiny apparent separation between an orbiting planet and its parent star nor block out its billion-times-brighter glare. Not even the largest of today's Earth-based telescopes, nor the orbiting Hubble, can detect another Earth circling Proxima Centauri.

For much of the 20th century the search for worlds around other stars seemed hopeless. But in the 1980s clever astronomers realized that

PLANETS BEYOND OUR SOLAR SYSTEM

Since 1995, scientists have discovered more than 300 extrasolar planets in our galaxy, most of them within 300 light-years of Earth. We describe the various types of planets on the pages that follow. For the latest planet count and an atlas of these new worlds, go to: http://planetquest.jpl.nasa.gov/index.cfm.

Total exoplanets	303
Stars with planets	259
Gas Giants	200
Hot Jupiters	76
Pulsar Planets	4
Earthlike Planets	0

there just might be a way. They calculated that the gravity of an unseen planet, minute as it was, would tug at its star, first from one side of its orbit, then the other. (Jupiter causes the sun to wobble back and forth every six years by about 461,000 miles.) This slight motion of the parent star can be detected using the Doppler effect: Its motion toward or away from us causes a tiny shift in the starlight's wavelength. If the star moves toward us, its light shifts to shorter, or bluer, wavelengths. When the planet tugs the star away, its light becomes more reddish. We can measure this Doppler shift using a telescope feeding the starlight to a precision spectrometer. The wavelength shift tells us how fast the star is moving relative to us, and thus the mass of the unseen companion. By timing the cycle of motion toward and away from us, we can also determine the orbital period (and thus distance) of the invisible planet from its star.

The 1995 discovery of 51 Pegasi by this method led to a surge in discoveries of planets beyond our solar system as more astronomers took up the exacting challenge. Today we know of 303 known extrasolar planets circling 259 different stars. As instrument sensitivity and telescopes improved, astronomers eventually were able to turn to methods of detection other than the Doppler effect.

Detecting the reflected light of a planet around a star is like trying to discern the glow of a firefly in the beam of a high-intensity searchlight.

Collapsing Solar Nebula The starting point for any new solar system—including our own—is the collapse of a cold cloud of gas and dust mixed with debris scattered by nearby, dying stars. The cloud's collapse accelerates under its own gravity. In less than 100,000 years, a new star begins burning hydrogen at its core in the process of nuclear fusion. Somewhere in the surrounding disk, planets may be forming.

SO CLOSE, SO FAR

just spent a marvelous night pass upstairs . . . sweeping across the Indian Ocean, out across the Philippines. If I can describe it: I saw more stars than I've ever seen from the ground. The stars were brilliant in the Southern Hemisphere—the Southern Cross was skimming the southern horizon the entire time, right down through the airglow. . . . Thunderstorms over Southeast Asia, the spangled cloth of the Milky Way stretching from my lower left to my upper right in my field of view toward the Magellanic Clouds . . . An occasional meteor below. Saw the city of Bangkok all lit up. Just a marvelous sky. I [tell] school audiences that if you go out on a dark night you'll see as many stars as you can from space, but it's not true. There are more stars in space than you can see from the ground. And they're all more bright—the brighter stars don't stand out as much. Part of a marvelous night scene from the orbiter Columbia. . . ."

Reading this passage spoken into my onboard recorder during my third orbital mission in late 1996, I can almost re-create the feeling of floating into that star field. My crewmates and I would marvel at the clouds of stars filling the sky above our dark Earth's horizon, softly lit by star- or moonlight. We were 200 miles up, but we wanted to go beyond: To know what was out there among the thousands of glowing stars in the black velvet sky. Our imaginations couldn't help but race out into that fantastic universe.   —Tom Jones

Not Even Rockets An astronaut can seemingly touch the stars, but a vast gulf separates us from the planets and the alien suns beyond. The moon is 240,000 miles distant: three days away using chemical rockets. Earth-approaching asteroids can be reached in round-trip voyages lasting three to four months. At closest approach, Mars is about 35 million miles away, nearly 150 times farther than the moon. But our rockets cannot track that straight line. We must use a curving, fuel-efficient trajectory that takes anywhere from six to nine months. A round trip, allowing for realignment of Earth and Mars, would take at least 2.5 years. Reaching Pluto, 3.7 billion miles from the sun, takes at least a ten-year cruise in deep space—one way. Proxima Centauri, 4.22 light-years away, is in another league altogether. A conventional spacecraft would require more than 73,000 years to reach this nearest star.

Seeing the Invisible

SEE ALSO: *Playing Hard to Find* 194

ONCE THE FIRST EXTRASOLAR PLANET was found, astronomers raced to find more. The initial surge in discoveries came using the Doppler method, since teams of researchers knew how to build the sensitive spectrometers required to detect the wavelength shift representing a wobble of a few feet per second. But astronomers soon found other viable methods of detection. Some planets pass in front of their parent star as they orbit, diminishing the star's brightness ever so slightly. We can detect this, and through repeated eclipses, this transit method can measure the planet's orbital period. As instruments become more precise, we can also detect a star's planet-caused wobble compared to more distant background stars. The size of the wobble detected by this astrometric technique enables us to measure the mass and orbit of the unseen planet.

Astronomers have detected a few exoplanets (another term for extrasolar planets) when they were magnified, in a sense, by the gravity of a random star passing between the planet and Earth. The nearer star's gravity bends and focuses the parent star's light, and the planet's presence creates a tiny irregularity in that pattern. The asymmetry of the pattern reveals the presence of an exoplanet. This microlensing technique can't be widely used, since the alignments occur so rarely. One other method has been useful for finding planets around neutron stars, the incredibly dense remnants of supernova explosions. This pulsar method works by noting slight variations in the pulsar's rotation as the unseen planet orbits the neutron star. These exoplanets must be cinders of worlds that once circled their star before it detonated in a catastrophic explosion.

Astronomers hope within the coming decade to obtain direct images of exoplanets, using techniques that eliminate the glare from the parent star. One method is to design the telescope to precisely mask the star's incoming light while permitting the dim reflected glow from the planet to shine through. A second technique—the interferometry method—combines the light from two telescopes but introduces into one a slight shift in the phase of the incoming starlight's wave pattern, cutting out the light from the parent star.

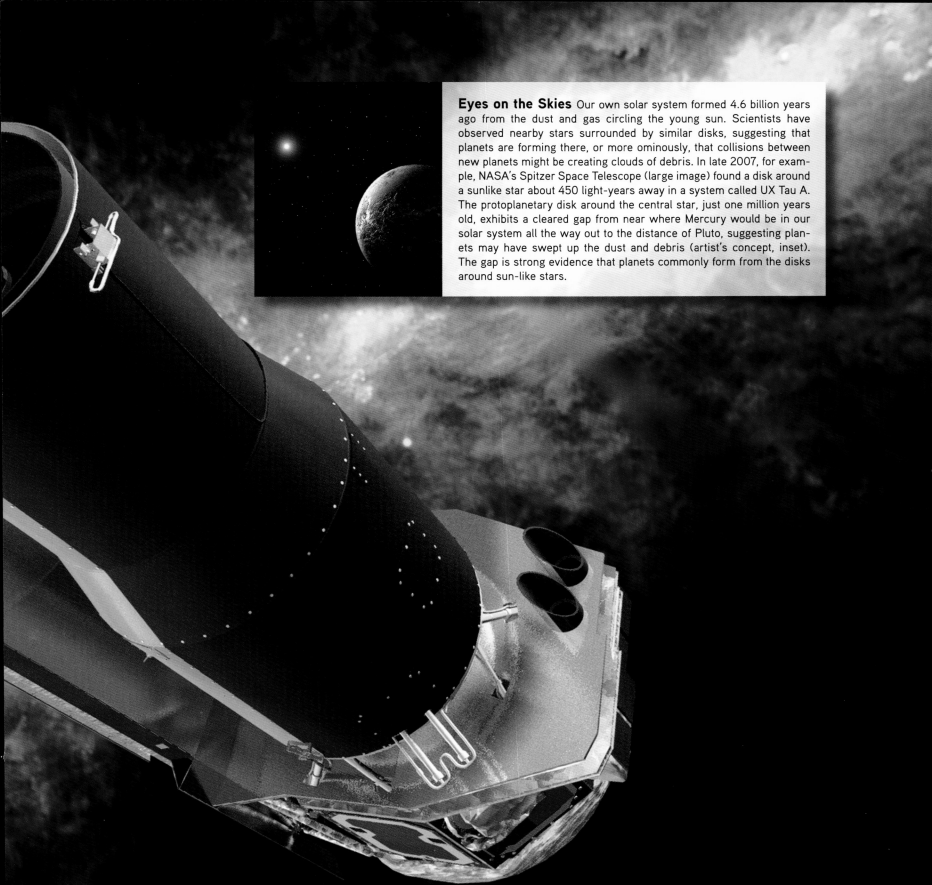

Eyes on the Skies Our own solar system formed 4.6 billion years ago from the dust and gas circling the young sun. Scientists have observed nearby stars surrounded by similar disks, suggesting that planets are forming there, or more ominously, that collisions between new planets might be creating clouds of debris. In late 2007, for example, NASA's Spitzer Space Telescope (large image) found a disk around a sunlike star about 450 light-years away in a system called UX Tau A. The protoplanetary disk around the central star, just one million years old, exhibits a cleared gap from near where Mercury would be in our solar system all the way out to the distance of Pluto, suggesting planets may have swept up the dust and debris (artist's concept, inset). The gap is strong evidence that planets commonly form from the disks around sun-like stars.

A Solar System's Birth Four hundred light-years away in the Pleiades (large image), in 2007 the Spitzer telescope found that one of the cluster's stars, HD 23514, is surrounded by a dense disk of hot dust particles (inset). The heat radiation from the dust, measured by Hawaii's Gemini telescope, was far higher than expected: The disk contains hundreds of thousands of times as much dust as our solar system. Monster collisions between primordial planets may be creating this debris cloud—the young, rocky planets around HD 23514 appear to be growing, in just the way we think Venus, Earth, and Mars grew, by absorbing smaller protoplanets. In the Pleiades we may have spotted another solar system in the violent throes of its birth.

The Planetary Zoo

THE DENIZENS OF OUR own solar system exhibit plenty of individuality, but what we see out there in the Milky Way is truly exotic. We have seen broiling-hot Jupiter-size worlds that race around their suns in just over a day, frozen planets that circle at the edge of the dim warmth thrown off by red dwarf stars, and chaotic planetary systems whose bodies career into each other in catastrophic collisions. The exoplanets discovered so far fall into six main categories.

Some are ultrahot Jupiters, who circle their stars at a blistering pace, once around in less than two days. About six of these, all more than half again as massive as Jupiter, are known.

The second variety we commonly see are called hot Jupiters, because they too circle much nearer to their sun than our Jupiter does. We know of about 50 of these, which revolve around their star in circular orbits with periods of two to seven days.

The majority of the exoplanets discovered are so-called eccentric giants, with an average mass of about five Jupiters. These worlds swing about their stars in highly elliptical orbits. About 7 percent of sunlike stars appear to be accompanied by these massive, looping planets.

Closer in size to our own solar system's gas goliaths are worlds termed long-period giants. These worlds are similar in size and solar distance to Jupiter or Uranus, but the dozen years or more they take to orbit (and thus tug on their stars) means that we are just beginning to notice the wobbles they impose on their suns. Observers hope that their large orbits will eventually make it easier for our instruments to block the glare of their central stars effectively and detect the planets directly.

Another species of exoplanets similar to our sun's family are the hot Neptunes, 5 to 30 times Earth's mass. But there the resemblance to our

Hot Jupiters (artist's concept, below) orbit their parent stars in a few days, close enough for their regular gravitational tugs to be detected from Earth.

solar system ends. These objects circle their low-mass stars so closely that they complete a revolution in two or three days.

Of the 303 exoplanets detected, the most numerous have a mass about 1.5 times that of Jupiter. But we are most curious, of course, about finding much smaller worlds, resembling our own Earth. These terrestrial planets, those half to twice the diameter of Earth and anywhere from a tenth to ten times its mass, are too small to exert much of a pull on their stars, and so are beyond the current limit of our telescopes' sensitivity. A few Earth-size worlds have been located around pulsars, but the supernova event that created the parent neutron star would have permanently sterilized these worlds.

Our search for Earth-size worlds continues. Beginning in 2005, astronomers found the smallest planets yet around two curious red dwarf stars. Red dwarves are very common: They make up 75 percent of the stars in our galaxy, and comprise 80 of the nearest 100 stars in our sun's neighborhood. Gliese 876 (located in the constellation Aquarius) is just 15 light-years away, and through the Doppler method astronomers found an exoplanet circling it that has just 7.5 times Earth's mass. But Gliese 876d, as it is known, circles its red dwarf parent at a blistering distance of two million miles; its sun appears 24 times as large in the sky as does our moon. Making a complete orbit in just 46 hours, its temperature ranges from 400°F to 700°F, and no one knows if it is a sizzling rock or a steaming, gas-shrouded world.

A second promising Earth analog was found in 2007. Gliese 581 is another red dwarf, 20.5 light-years away in the constellation Libra, one-third the mass and just one-fiftieth the brightness of our sun. The third exoplanet discovered around Gliese 581 is only half again as wide as our planet, with a mass just five times Earth's. Although it orbits 14 times closer to its sun than does Earth, the dim output from its parent means that this world might be in the "sweet spot"—where temperatures allow the existence of liquid water.

The holy grail of exoplanet searches is an Earthlike planet (artists' concept, right) in the habitable zone around its star.

In the Zone

SEE ALSO: *The Rise of Life* 170

LTHOUGH ONLY A HANDFUL of planets in Earth's size range are known, those numbers are sure to increase as our telescopes become more capable. We're interested in finding rocky worlds like our own where liquid water might be stable at the surface—one of our basic prerequisites for life. That region around each star where temperatures are moderate enough to allow liquid water is called the habitable zone.

As we learned in chapter 8, the habitable zone for sunlike stars is where liquid water is stable on a planetary surface, generally a zone stretching between the orbits of Mars and Venus. At Gliese 581, exoplanet 581c is slightly too close to its star to be habitable, but 581d is within the outer edge of the star's calculated habitable zone. Thus, Gliese 581d has the potential for hosting liquid water. Excitement over finding this temperate super-Earth is somewhat dampened by its eccentric, star-hugging orbit, so close that gravity has permanently locked one hemisphere facing its sun, creating surface temperature extremes between the sunlit and shadowed sides. We have yet to find a world with the same moderate conditions our own Earth experiences.

But improvements in our detection technology within 20 years should enable the discovery of more candidate Earthlike exoplanets. Once we have found such worlds within their stars' habitable zones, the next phase of our search begins.

We will use the largest ground-based telescopes and a new generation of space-based detectors to probe their compositions by measuring the telltale spectra of light reflected from these pseudo-Earths. The pattern of starlight reflected and absorbed by their atmospheres can yield crucial information about the gases there: We know that molecular oxygen, ozone, and water create strong absorption features that our current spectrometers can detect. We're most interested in knowing whether an Earth-size world has free oxygen and water vapor, both strongly suggestive that the planet is habitable and that some process there is pumping free oxygen into the atmosphere. That something might just be life.

Life's Abundance Bora Bora, in the Pacific's Society Islands, exemplifies the hospitable climate made possible by our planet's location and unique combination of life-giving elements. Soon, large ground-based telescopes and space-based detectors will probe newly discovered Earth-like worlds. Will we find one in the zone of habitability, located just the right distance from its star, with the chemical prerequisites for life? What would "life" on another "Earth" look like? Our curiosity drives the urge to explore space for answers.

Finding Other Earths

LTHOUGH WITH OUR PRESENT technology we cannot detect a planet the size of Earth around even the closest stars, that limitation will fall to improved telescope designs, especially those based in space. The Hubble Space Telescope has been used to detect planet-induced stellar wobbles, and was the first to detect the constituents of an exoplanet's atmosphere. NASA's Kepler mission will orbit a 3.2-foot-diameter telescope in 2009 whose sensitive photometer will detect Earth-size planets transiting their stars. A follow-on mission, SIM PlanetQuest, will use interferometry to measure the precise distances and positions of the Milky Way's stars, with enough accuracy to note if exoplanets slightly larger than Earth are tugging their stellar parents.

A long-term NASA goal is to launch a large third-generation space telescope called Terrestrial Planet Finder. The two Terrestrial Planet Finder observatories will be capable of detecting and characterizing Earthlike planets around as many as 200 stars up to 45 light-years away. The project will look for the atmospheric signatures (such as water, carbon dioxide, and ozone) of habitable or even "inhabited" planets (where life—not necessarily intelligent—indeed exists). TPF's satellites will use both the blocking coronagraph or interferometry technique to cut the

Searching for Earths New instruments, some bound for the biggest telescopes on Mauna Kea on Hawaii (large image), some to reside in space, aim to measure the atmospheric composition of extrasolar "Earths." The right mix of gases could point to the presence of biology. Free oxygen and water could indicate that photosynthesis or a similar process is active. Methane and nitrous oxide are also life-connected compounds that would be sure to spur our efforts to explore other Earths.

glare from the parent star's light and directly image exoplanets. This ambitious observing mission, probably 20 years or more in the future, will give us our first in-depth look at Earthlike worlds discovered in the coming decades by ground- and space-based searches.

In orbit, astronauts often seek out the night sky during their 45-minute passes through Earth's shadow. Floating in the cabin, faces close to the window, they can see the spangled belt of the Milky Way, seemingly adrift in a sea of stars. What unknown worlds in their millions exist out there? Is anyone staring back? If so, what might their world be like?

Back on the ground, like so many humans, we have been fascinated and educated by examining the many geologic features of our solar system to be found and studied right here on Earth. Today, our ground-based telescopes are fully engaged in the search for other Earths. When we find them, our quest to understand their surfaces and suitability for life will again be based on what we've learned on Earth and its closest planetary neighbors. Whether we use our own eyes and hands or send our robot probes to extend our senses, making sense of what we find will always bring us back to this island Earth, our textbook for knowing other worlds.

Further Reading

Chapter 1

Beatty, J.K., C.C. Petersen, and A. Chaiken, eds. *The New Solar System*. Cambridge, MA: Cambridge University Press, 1998.

Chapman, Clark R. *Planets of Rock and Ice*. New York: Scribners, 1982.

Jones, Tom. *Sky Walking: An Astronaut's Memoir*, New York: Smithsonian-Collins, 2006.

New Frontiers in the Solar System: An Integrated Exploration Strategy, Solar System Exploration Survey. National Research Council, 2003.

McFadden, L.A., P.R. Weissman, and T.V. Johnson eds. *Encyclopedia of the Solar System*, 2nd Edition. San Diego: Academic Press, 2007.

Stern, S.A. ed., *Our Worlds: The Magnetism and Thrill of Planetary Exploration*. Cambridge: Cambridge University Press, 1999.
Gateway to Astronaut Photography of Earth
http://eol.jsc.nasa.gov/sseop/clickmap/

Earth Observatory
http://earthobservatory.nasa.gov/

Visible Earth
http://visibleearth.nasa.gov/

Cities Collection
http://eol.jsc.nasa.gov/cities/

SIR-C X-SAR images
http://www.jpl.nasa.gov/radar/sircxsar/

Shuttle Radar Topography Mission
http://www2.jpl.nasa.gov/srtm/

General NASA information
www.nasa.gov

European Space Agency
http://www.esa.int/esaCP/index.html

Apollo Lunar Surface Journal
http://history.nasa.gov/alsj/

Nine Planets—solar system information
http://nineplanets.org/

NASA images
http://photojournal.jpl.nasa.gov

Mars exploration program
http://mars.jpl.nasa.gov/

Chapter 2

Schmitt, Harrison H. *Return to the Moon*. Copernicus Books, 2006.

Spudis, Paul D. *The Once and Future Moon*. Washington, D.C.: Smithsonian Press, 1996.

Association of Space Explorers–Near Earth Object Committee
http://space-explorers.org/committees/NEO/neo.html

NASA NEO Program Office
http://neo.jpl.nasa.gov/index.html

Yellowstone Volcano Observatory
http://volcanoes.usgs.gov/yvo/index.html

Supervolcanoes—Scientific American
http://www.sciam.com/article.cfm?id=the-secrets-of-supervolca

Deep Impact comet mission
http://solarsystem.nasa.gov/deepimpact/index.cfm

NEAR Shoemaker—mission to Eros
http://near.jhuapl.edu/

Hayabusa asteroid mission
(Japanese Space Agency JAXA)
http://www.isas.ac.jp/e/enterp/missions/hayabusa/index.shtml

Dawn asteroid mission
http://dawn.jpl.nasa.gov/

Chapter 3

Cooper, H.S.F. *The Evening Star*. New York: Farrar, Straus, Giroux, 1993.

Grinspoon, D.H., *Venus Revealed*. New York: Basic Books, 1997.

Hamblin, W.K. and E.H. Christiansen, *Earth's Dynamic Systems* 10th Edition. New York: Prentice Hall, 2003.

Hartmann, W.K. *A Traveler's Guide to Mars*. Workman Publishing, 2003.

Squyres, S. *Roving Mars: Spirit, Opportunity, and the Exploration of the Red Planet*. New York: Hyperion, 2005.

Geology, earthquakes, volcanoes
http://www.usgs.gov/

Chapter 4

Lopes, Rosaly, and M.C. Carroll *Alien Volcanoes*. Baltimore: Johns Hopkins University Press. 2008.

Lopes, R.M.C. and T.K.P. Gregg eds. *Volcanic Worlds: Exploring the Solar System's Volcanoes*. Berlin: Springer-Verlag, 2004.

Sigurdsson, H., ed. *Encyclopedia of Volcanoes*. San Diego, Academic Press, 2000.

Volcano World—University of North Dakota
http://volcano.und.edu/

Chapter 5

Mars Global Surveyor
http://mars.jpl.nasa.gov/mgs/

Mars Phoenix Lander Mission
http://phoenix.lpl.arizona.edu/

Cassini Imaging Home Page
http://ciclops.org/index.php

Chapter 6

Benn, D.I. and Evans, D.J.A. *Glaciers and Glaciation*. London: Arnold, 1998.

Lorenz, R. and J. Mitton, *Titan Unveiled: Saturn's Mysterious Moon Explored*. Cambridge MA: Cambridge University Press, 2008.

National Snow and Ice Data Center
http://nsidc.org/glaciers/

Panel on Climate Change
http://www.ipcc.ch/Intergovernmental

Chapter 7

Bell, Jim. *Postcards from Mars*. New York: Dutton, 2007.

CRISM spectrometer on Mars Reconnaissance Orbiter
http://crism.jhuapl.edu/index.php

Mars Reconnaissance Orbiter
http://mars.jpl.nasa.gov/mro/

University of Arizona Lunar and Planetary Laboratory
http://www.lpl.arizona.edu/

Chapter 8

Grinspoon, D. *Lonely Planets: The Natural Philosophy of Alien Life*. New York: Harper Collins, 2003.

Lunine, J.I. *Earth: Evolution of a Habitable World*. Cambridge MA: Cambridge University Press, 1999.

Committee on an Astrobiology Strategy for the Exploration of Mars, National Research

Council, *An Astrobiological Strategy for the Exploration of Mars,* Washington. D.C.: 2007.

Chapter 9

Casoli, Fabienne, and Encrenaz, T. *The New Worlds: Extrasolar Planets*. New York: Springer, 2007.

McSween, Harry Y. *Fanfare for Earth: The Origin of Our Planet and Life*. New York: St. Martin's Press, 1997.

PlanetQuest—the search for other worlds
http://planetquest.jpl.nasa.gov/

Terrestrial Planet Finder—searching out other Earths
http://planetquest.jpl.nasa.gov/TPF-C/tpf-C_index.cfm

University of Pittsburgh–Planet Search
http://www.pitt.edu/~aobsvtry/index.html

Kepler mission—search for habitable planets
http://kepler.nasa.gov/

Search for life
http://astrobiology.nasa.gov/

Index

Boldface indicates illustrations.

About the Authors

TOM JONES is a planetary scientist, author, speaker, and former NASA astronaut. He flew on four space shuttle missions and led three space walks to help his crew install the U.S. Destiny lab, centerpiece of the International Space Station. He is the author of *Sky Walking: An Astronaut's Memoir* (Harper Collins, 2006) and *Hell Hawks!* (with Robert F. Dorr, Zenith Press, 2008). He writes frequently for Aerospace America and Air & Space Smithsonian, and is a regular on-air contributor to Fox News' spaceflight coverage. A board member of the Association of Space Explorers, he writes, consults, and speaks from Houston, Texas. Visit www.AstronautTomJones.com.

ELLEN STOFAN is a planetary geologist who has conducted research on the geology of volcanic and tectonic features on Venus, Mars, Titan, and Earth. While an employee of NASA's Jet Propulsion Laboratory (JPL) she was the chief scientist on NASA's New Millennium Program and the deputy project scientist on the Magellan Mission to Venus. She is currently a senior research scientist at Proxemy Research and an honorary professor of earth sciences at University College London. She lives with her husband and three children on a farm in Virginia.

Acknowledgments

PLANETOLOGY WOULD NOT have been written without the constant support and understanding of the authors' spouses: Liz Jones and Tim Dunn. They stoically tolerated the authors' months of research and writing. Long before this writing project began, Liz and Tim encouraged (despite our extended absences) countless field geology trips, planetary science conferences, shuttle training expeditions, mission simulations, and even many weeks spent off this planet. Their selflessness truly made this book possible. Our children, too—Annie and Bryce Jones, and Ryan, Emily, and Sarah Dunn—deserve our thanks.

The authors thank their literary agent, Deborah C. Grosvenor, at Kneerim & Williams at Fish and Richardson, for her encouragement, patience, and steady advocacy of this project.

We could not have been more delighted to work with National Geographic Books and its team of editors, designers, and imagery experts. We thank our editors, Lisa Thomas and Barbara Seeber; our graphics and image expert, Erin Benit; designer and layout specialist Cameron Zotter; and our valued assistant editors, Olivia Garnett and Judith Klein.

The authors are grateful for the encouragement and advice offered by their colleagues in planetary science, especially Jonathan Lunine, Cindy Evans, Faith Vilas, Mark Robinson, John Guest, Steve Anderson and Sue Smrekar. Any errors of fact, interpretation, or explanation are wholly the responsibility of the authors.

We dedicate *Planetology* to those explorers who have devoted their careers to pushing back the frontiers of our understanding of the solar system. The amazing scientific findings described in this book are due to the hard work of the Earth and planetary science communities. We especially remember those scientists on Earth and those space travelers who have risked and tragically lost their lives in our efforts to discover our place in the cosmos. As we send our robots to the farthest reaches of the solar system, launch giant telescopes into space to search for planets around other stars, and one day, stand on the bedrock of distant worlds, we will not forget the explorers of Apollo 1, Soyuz 1, Soyuz 11, *Challenger*, and *Columbia*.

Illustrations Credits

i, NASA/JPL/University of Arizona; ii-iii, NASA and The Hubble Heritage Team (STScI/AURA); iv-v, Image Science and Analysis Laboratory, NASA-Johnson Space Center. "The Gateway to Astronaut Photography of Earth."; 2-3, NASA images by Reto Stöckli, based on data from NASA and NOAA; 4-5, NASA/JPL-Caltech/University of Arizona/Cornell/Ohio State University; 5 (LO), Courtesy of NASA's National Space Science Data Center; 5 (UP), Landsat imagery courtesy of NASA Goddard Space Flight Center and U.S. Geological Survey; 6-7, Bates Littlehales; 7 (UP), Image Science and Analysis Laboratory, NASA-Johnson Space Center. "The Gateway to Astronaut Photography of Earth."; 8-9, NASA/JPL; 10, NASA/JPL; 11 (UP), NASA/JPL/Space Science Institute; 11 (LO), ESA/DLR/FU Berlin (G. Neukum); 12-13, GeoEye; 13 (UP), Image Science and Analysis Laboratory, NASA-Johnson Space Center. "The Gateway to Astronaut Photography of Earth."; 13 (CTR), Landsat imagery courtesy of NASA Goddard Space Flight Center and U.S. Geological Survey; 13 (LO), ESA; 14-15, NASA/JPL/University of Arizona; 15, Gordon Wiltsie/NG Image Collection; 16-17, Robert W. Madden; 17, Thomas Jones; 18, Courtesy of the NAIC - Arecibo Observatory, a facility of the NSF; 19, ESA; 19 (UP), NASA/JPL; 20 (UP LE), Wikipedia; 20 (UP RT), Courtesy of NASA's National Space Science Data Center; 20 (LO LE), Wikipedia; 20 (LO CTR), Wikipedia; 20 (LO RT), NASA; 21 (UP LE), NASA; 21 (UP RT), McREL for NASA; 21 (LO LE), Courtesy of NASA's National Space Science Data Center; 21 (LO RT), NASA; 22-23, NASA/JPL/Space Science Institute; 22 (LE), NASA/JPL/Space Science Institute; 24-25, Adriel Heisey; 26, Courtesy of The Macovich Collection; 27, Landsat imagery courtesy of NASA Goddard Space Flight Center and U.S. Geological Survey; 27 (UP LE), NASA; 28-29, NASA; 29, Wikipedia; 30-31, Harman Smith and Laura Generosa (nee Berwin), graphic artists and contractors to NASA's Jet Propulsion Laboratory.; 31 (UP), T. A. Rector (University of Alaska Anchorage), Z. Levay and L.Frattare (Space Telescope Science Institute) and WIYN/NOAO/AURA/NSF; 31 (CTR), Fred Bruenjes; 31 (LO), NASA/JPL; 32, 33, NASA/GSFC/METI/ERSDAC/JAROS and U.S./Japan ASTER Science Team; 33 (RT), NASA/JPL/Space Science Institute; 34-35, NASA/Goddard Space Flight Center Scientific Visualization Studio; 34 (UP LE), DigitalGlobe and Satellite Imaging Corporation; 34 (UP RT), NASA; 34 (LO), NASA/JPL; 35 (UP LE), Jiri Moucka/iStockphoto.com; 35 (UP RT), NASA; 35 (CTR), NASA; 35 (LO LE), NASA/GSFC/METI/ERSDAC/JAROS and U.S./Japan ASTER Science Team; 35 (LO CTR), NASA; 35 (LO RT), NASA; 36, NASA/JPL/Space Science Institute; 36-37, NASA/Johns Hopkins University Applied Physics Laboratory/Carnegie Institution of Washington; 37, NASA John W. Young; 38, 39, Courtesy of Fahad Sulehria, http://www.novacelestia.com; 40-41, Courtesy of JAXA, ISAS; 41, Tunc Tezel; 41 (RT), Ali Jarekji/Reuters/CORBIS; 42-43, NASA/JPL/Space Science Institute, David Seal; 43 (RT), H. Hammel, MIT and NASA; 43 (UP LE), NASA/Calar Alto Observatory; 44-45, NASA; 45 (RT), Galileo Project, Brown University, JPL, NASA; 46, Pixeldust Studios; 47, Jacques Descloitres, MODIS Land Science Team; 47 (LO RT), NASA/GSFC/METI/ERSDAC/JAROS and U.S./Japan ASTER Science Team; 48, Zhifeng Wang/iStockphoto.com; 48-49, Frans Lanting; 49, NASA/JPL; 50-51, Yann Arthus-Bertrand/CORBIS; 51, Tim Dunn/Mt. Etna, Sicily/Ellen Stofan, Sarah Dunn; 52, NASA/JPL; 53, Bartosz Wardzinski/Shut-

terstock; 54, Paul A. Souders/CORBIS; 54-54, NASA/Johns Hopkins University Applied Physics Laboratory (or NASA/JHUAPL); 55, Courtesy of Palomar Observatory, California Institute of Technology; 56-57, Winfield I. Parks, Jr.; 58-59, Jaime Quintero; 59, NASA; 60, NASA/JPL/Space Science Institute; 60 (LO), NASA/JPL/NIMA; 61 (LO), Susan Sanford; 61 (CTR), Susan Sanford; 61 (UP), Susan Sanford; 62, NASA; 62-63, NASA/JPL/ASU; 64-65, NASA; 65, Sisse Brimberg; 66, NASA/JPL/USGS; 67, NASA/JPL/Malin Space Science Systems; 67 (LO), ESA; 68, NASA/JPL/Space Science Institute; 69, NASA/JPL; 70, NASA/JPL/University of Arizona; 71, NASA/JPL/University of Arizona; 71 (CTR), NASA/Johns Hopkins University Applied Physics Laboratory/Carnegie Institution of Washington; 72-73, James L. Stanfield; 73, George F. Mobley; 75, NASA/JPL; 75 (LO), Courtesy of GeoEye; 76-77, Ken Straiton/CORBIS; 77, AP/Wide World Photos/Oded Balilty; 78-79, Peter Carsten/NG Image Collection; 80, Dorling Kindersley/Getty Images; 81, NASA; 81 (LO), NASA/JPL Magellan/E.R. Stofan; 82-83, Jaime Quintero; 83 (LE), Jacques Descloitres, MODIS Rapid Response Team, NASA/GSFC; 83 (RT), NASA/JPL/USGS; 84-85, Cheryl Nuss; 85, O. Louis Mazzatenta; 86-87, NASA; 87, NASA; 88, NASA/JPL; 88 (LO), James L. Amos; 89, NASA/USGS; 90, NASA image created by Jesse Allen, Earth Observatory, using data provided courtesy of NASA/GSFC/MITI/ERSDAC/JAROS, and U.S./Japan ASTER Science Team.; 91, NASA/JPL/University of Arizona; 91 (RT), NASA; 92, NASA; 93, NASA/JPL/ASU; 93 (LO RT), NASA/JPL/University of Arizona; 94-95, NASA; 95, NASA/JPL; 96, NASA/JPL/USGS; 97, NASA/JPL/Space Science Institute; 97 (LO), NASA/JPL Cassini-Huygens Radar Mapper/E.R. Stofan; 98-99, NASA/JPL/DLR; 100-101, Emory Kristof; 101, Sarah Leen; 102-103, Image courtesy Jacques Descloitres, MODIS Land Rapid Response Team at NASA GSFC; 104, Marlene DeGrood/iStockphoto.com; 105, Paul Chesley; 105 (CTR), NASA/JPL/USGS; 106-107, NASA image created by Jesse Allen, using data provided by the University of Maryland's Global Land Cover Facility; 107, NASA image created by Jesse Allen, Earth Observatory, using data provided courtesy of NASA/GSFC/MITI/ERSDAC/JAROS, and U.S./Japan ASTER Science Team; 108, Walter M. Edwards; 109, NASA images created by Jesse Allen, Earth Observatory, using data provided courtesy of the Landsat Project Science Office; 109 (LO RT), Andrea Booher/FEMA Photo; 110, ESA, NASA, Descent Imager/Spectral Radiometer Team (LPL); 111, ESA/DLR/FU (G. Neukum); 112, NASA/JPL/ASU; 113, NASA/GSFC/METI/ERSDAC/JAROS and U.S./Japan ASTER Science Team; 114, ESA/ DLR/FU Berlin (G. Neukum); 114-115, Joel Sartore; 115, Ralph Gray; 116, NASA/JPL/Malin Space Science Systems; 117, Brenda Beitler, University of Utah; 117 (UP RT), NASA/JPL/Cornell/USGS; 118-119, South Florida Water Management District; 119, South Florida Water Management District; 119 (UP LE), NASA/JPL/USGS; 120-121, Frans Lanting; 121, NASA; 122-123, NASA; 123, Photographer's Mate 2nd Class Philip A. McDaniel; 124-125, Maria Stenzel; 126, Risteski Goce/Shutterstock; 127, Brian Skerry; 127 (RT), Frank Hurley/CORBIS; 128-129, ESA/AOES Medialab; 130, EESA/DLR/FU Berlin (G. Neukum); 130-131, NASA/JPL/Malin Space Science Systems; 131, NASA/JPL/MSSS; 132-133, NASA/JPL/University of Arizona; 133, Mikhail Pogosov/Shutterstock; 134, South Tyrol Museum of Archeology (Bolzano Italy); 135, IKONOS satellite

image courtesy GeoEye. Image interpretation courtesy Ted Scambos, National Snow and Ice Data Center; and Tad Pfeffer, Institute of Arctic and Alpine Research.; 135 (UP RT), S. Greg Panosian/iStockphoto.com; 136-137, James L. Amos; 136 (LO), NASA/JPL/Malin Space Science Systems; 138-139, ESA/DLR/FU Berlin (G. Neukum); 139, Maria Stenzel; 140, Map Compilation: Technische Universität Berlin, 2006; Image Data: ESA / DLR / FU Berlin (G. Neukum); 141, Oddur Sigurðsson; 141 (UP RT), NASA/JPL/Arizona State University; 142-143, Jon Larson/iStockphoto.com; 142 (LE), NASA; 144-145, NASA/JPL; 145 (UP), NASA/Goddard Space Flight Center Scientific Visualization Studio; 145 (LO), NASA/Goddard Space Flight Center Scientific Visualization Studio; 146-147, Landsat imagery courtesy of NASA Goddard Space Flight Center and U.S. Geological Survey; 148, National Weather Service; 149, NASA; 149 (LO), ESA/VIRTIS/INAF-IASF/Obs. de Paris-LESIA; 150-151, NASA; 151, NASA/JPL/Space Science Institute; 152, George Steinmetz; 153, ESA/DLR/FU Berlin (G. Neukum); 153 (UP RT), NOAA George E. Marsh Album; 154-155, NASA/JPL/Cornell; 155, Ira Block/NG Image Collection; 156, AP/Wide World Photos/Laura Rauch; 157, NASA/Malin Space Science Systems; 157 (UP RT), NASA/JPL/MSSS; 158-159, Robert Sisson; 160, NASA/JPL/Malin Space Science Systems; 160-161, NASA/JPL-Caltech/Cornell; 161, NASA/JPL/University of Arizona; 162, NASA/JPL/Malin Space Science Systems; 163, Thomas J. Abercrombie; 163 (LO RT), NASA; 164-165, NASA/JPL; 165, NASA/JPL; 166-167, National Oceanic and Atmospheric Administration/Department of Commerce; 167 (CTR), George Steinmetz; 168-169, Robert B. Haas; 170, Wikipedia; 171, The Natural History Museum/Alamy Ltd; 172, AP/Wide World Photos; 173, Nemo Ranjet/Science Photo Library; 174-175, National Oceanic and Atmospheric Administration/Department of Commerce; 175, Frank Lane Picture Agency/CORBIS; 176-177, Seth Shostak; 177 (UP), NASA; 177 (CTR), NASA/Johns Hopkins University Applied Physics Laboratory/Southwest Research Institute; 177 (LO), NASA/JPL/Space Science Institute; 178-179, ESA/DLR/FU Berlin (G. Neukum); 178, ESA/DLR/FU Berlin (G. Neukum); 179, Norbert Rosing/NG Image Collection; 180, NASA/JPL/University of Arizona; 181, Pacific Ring of Fire 2004 Expedition. NOAA Office of Ocean Exploration; Dr. Bob Embley, NOAA PMEL, Chief Scientist; 182-183, NASA/JPL; 183, Annie Griffiths Belt; 184-185, NASA/JPL-Caltech; 185, Michael Nichols, NGP; 186, Roger Ressmeyer/CORBIS; 187, ESA; 188-189 (LE), Louis Psihoyos/CORBIS; 188-189 (RT), Chris Newbert/Minden Pictures/NG Image Collection; 190-191, NASA, ESA, and The Hubble Heritage Team (STScI/AURA); 192, Stephen L. Alvarez; 192-193, NASA, ESA, S. Beckwith (STScI), and The Hubble Heritage Team (STScI/AURA); 193, MicroFUN Collaboration, CfA, National Science Foundation; 194, Mark Thiessen, NGP; 195, NASA/JPL-Caltech; 196-197, Image courtesy of the Image Science & Analysis Laboratory, NASA Johnson Space Center; 197, NASA; 198-199, NASA/JPL-Caltech/K. Su (Univ. of Ariz.) & NASA/JPL-Caltech; 199, NASA, ESA and G. Bacon (STScI); 200-201, NASA, ESA and AURA/Cal; 200 (UP LE), NASA/JPL-Caltech/J. Stauffer (SSC/Caltech); 202, ESA - C.Carreau; 203, Image courtesy of Trent Schindler and the National Science Foundation; 204-205, George F. Mobley; 206-207, NASA/ California Association for Research in Astronomy/ W. M. Keck Observatory/T. Wynne.

PLANETOLOGY

By Tom Jones and Ellen Stofan

PUBLISHED BY THE NATIONAL GEOGRAPHIC SOCIETY
John M. Fahey, Jr., President and Chief Executive Officer
Gilbert M. Grosvenor, Chairman of the Board
Tim T. Kelly, President, Global Media Group
John Q. Griffin, President, Publishing
Nina D. Hoffman, Executive Vice President;
    President, Book Publishing Group

PREPARED BY THE BOOK DIVISION
Kevin Mulroy, Senior Vice President and Publisher
Leah Bendavid-Val, Director of Photography Publishing
    and Illustrations
Marianne R. Koszorus, Director of Design
Barbara Brownell Grogan, Executive Editor
Elizabeth Newhouse, Director of Travel Publishing
Carl Mehler, Director of Maps

STAFF FOR THIS BOOK
Barbara Seeber and Lisa Thomas, Project Editors
Barbara Seeber, Text Editor
Erin Benit, Illustrations Editor
Peggy Archambault, Melissa Farris, and Cameron Zotter, Designers
Olivia Garnett, Associate Editor
Marshall Kiker, Illustrations Specialist
Mike Horenstein, Production Project Manager
Connie D. Binder, Indexer

Jennifer A. Thornton, Managing Editor
R. Gary Colbert, Production Director

MANUFACTURING AND QUALITY MANAGEMENT
Christopher A. Liedel, Chief Financial Officer
Phillip L. Schlosser, Vice President
Chris Brown, Technical Director
Nicole Elliott, Manager
Monika D. Lynde, Manager
Rachel Faulise, Manager

Founded in 1888, the National Geographic Society is one of the largest non-profit scientific and educational organizations in the world. It reaches more than 285 million people worldwide each month through its official journal, National Geographic, and its four other magazines; the National Geographic Channel; television documentaries; radio programs; films; books; videos and DVDs; maps; and interactive media. National Geographic has funded more than 8,000 scientific research projects and supports an education program combating geographic illiteracy.

For more information, please call 1-800-NGS LINE
(647-5463) or write to the following address:

National Geographic Society
1145 17th Street N.W.
Washington, D.C. 20036-4688 U.S.A.

Visit us online at www.nationalgeographic.com/books

For information about special discounts for bulk purchases, please contact National Geographic Books Special Sales: ngspecsales@ngs.org

For rights or permissions inquiries, please contact National Geographic Books Subsidiary Rights: ngbookrights@ngs.org

Library of Congress Cataloging-in-Publication Data

Jones, Tom, 1955 Jan. 22-
  Planetology : Unlocking the secrets of the solar system / [Tom Jones and Ellen Stofan].
     p. cm.
  Includes index.
  ISBN 978-1-4262-0121-9
  1. Planetology. 2. Planets--Geology 3. Earth--Origin. I. Stofan, Ellen Renee, 1961- II. Title.
  QB602.9.J66 2008
  523.4--dc22
                                            2008010726

ISBN: 978-1-4262-0121-9
Printed in China